Planewaves, Pseudopotentials and the LAPW Method

Second Edition

PLANEWAVES, PSEUDOPOTENTIALS AND THE LAPW METHOD

Second Edition

DAVID J. SINGH
Condensed Matter Sciences Division,
Oak Ridge National Laboratory
Oak Ridge, TN 37831-6032, U.S.A.

LARS NORDSTRÖM
Department of Physics, Uppsala University Uppsala, Sweden

David J. Singh
Naval Research Laboratory
Washington, D.C., USA

Lars Nordström
Uppsala University
Uppsala, Sweden

Planewaves, Pseudopotentials, and the LAPW Method, 2nd Ed

ISBN 13: 978-1-4419-3954-8

ISBN 10: 0-387-29684-0 (e-book)
eISBN 13: 978-0-387-29684-5

Printed on acid-free paper.

Printed in the United States of America.

springeronline.com

This book is dedicated to our wives, Nancy and Hedvig, whose love, patience and understanding made this work possible.

Contents

Dedication v

List of Figures ix

Preface xi

1. INTRODUCTION 1

2. DENSITY FUNCTIONAL THEORY AND METHODS 5
   - 2.1 Density Functional Theory 6
   - 2.2 Solution of the Single Particle Kohn-Sham Equations 10
   - 2.3 Self-Consistency in Density Functional Calculations 13
   - 2.4 Spin-Polarized Systems 16
   - 2.5 Non-Collinear Magnetism 18
   - 2.6 The LDA+U Method 20

3. PLANEWAVE PSEUDOPOTENTIAL METHODS 23
   - 3.1 Why Planewaves 24
   - 3.2 Pseudopotentials 26
   - 3.3 Introduction to the Car-Parrinello Method 36

4. INTRODUCTION TO THE LAPW METHOD 43
   - 4.1 The Augmented Planewave Method 43
   - 4.2 The LAPW Basis and its Properties 46
   - 4.3 Role of the Linearization Energies 49

5. NITTY-GRITTIES 53
   - 5.1 Representations of the Charge Density and Potential 53
   - 5.2 Solution of Poisson's Equation 57
   - 5.3 The Exchange Correlation Potential 60

5.4 Synthesis of the LAPW Basis Functions 62
5.5 Synthesis of the Hamiltonian and Overlap Matrices 67
5.6 Brillouin Zone Integration and the Fermi Energy 73
5.7 Computation of the Valence Charge Density 78
5.8 Core State Relaxation and Atomic Charge Densities 81
5.9 Multiple Windows and Local Orbital Extensions 83
5.10 The APW+LO Basis Set 88
5.11 Charge Density Mixing for Self-Consistency 90
5.12 Fixed Spin Moment Calculations 92
5.13 The Total Energy 94
5.14 Atomic Forces 95
5.15 Density Functional Perturbation Theory and Linear Response 99
5.16 Second Variational Treatment of Spin-Orbit Effects 102
5.17 Spin-Orbit with $p_{1/2}$ Local Orbitals 105
5.18 Iterative Diagonalization 106

6. CAR-PARRINELLO AND THE LAPW METHOD 107
6.1 Preliminaries 107
6.2 The Transformation of Goedecker and Maschke 109
6.3 The Transformation of Singh *et al.* 113
6.4 Status and Outlook 121

References 123

Index 133

List of Figures

2.1 Schematic flow-chart for self consistent density functional calculations. 14
3.1 Schematic illustration of the replacement of the all-electron wavefunction and core potential by a pseudo-wavefunction and pseudopotential. 25
3.2 Bachelet, Hamann, Schluter procedure for generating norm-conserving pseudopotentials. 28
4.1 The dual representation of the APW and LAPW methods. Stars and lattice harmonics are symmetrized planewaves and spherical harmonics used to represent the density and potential. 44
4.2 Procedure for setting the E_ℓ in the LAPW method. Note that the various ℓ components are to be considered separately. 51
5.1 Construction of the stars. 55
5.2 The construction of the lattice harmonics to represent the charge density and potential inside the LAPW spheres. 56
5.3 Pseudocharge method for solving the Poisson equation. 59
5.4 The exchange-correlation potential. 61
5.5 $\mathbf{A}^T$ and $\mathbf{R}$ (which can be precalculated outside the radial loop) are used to transform from lattice harmonics to real space and vice-versa. 62
5.6 The rotated coordinate system inside equivalent (*i.e.* symmetry related) atoms. Note that the representative atom is in the global frame. 66
5.7 Computation of $\tilde{V}_{PW}$ using fast Fourier transforms. 70
5.8 Evaluation of the non-spherical contribution to the Hamiltonian. The work inside the main loop is bypassed when the Gaunt coefficients (G) are zero. 73

5.9 The generation of special **k**-points and weights. The right side gives an example for a 2-dimensional square lattice with divisions of $1/4$ of the zone. 75

5.10 Computation of the interstitial charge density. The **k**-point sum is over the IBZ and the symmetrization is done by projecting onto stars. 79

5.11 The sphere charges. The symmetrization projects the charge in each atom onto the representative atom, so the the loop on n is on all atoms. 80

5.12 Conversion of extended core or atomic charge densities to the LAPW representation. 82

5.13 Example of windows with a semi-core state. The E_ℓ corresponding to the semi-core angular momentum is set low in the single window case. 84

5.14 Variation of a semi-core and a valence band with the linearization energy, E_ℓ. The dotted lines at ε_1 and ε_2 denote the true locations of the bands. 85

5.15 The fixed spin moment procedure. 93

5.16 The second variational procedure for treating spin-orbit. i and j are indices that run over spins and the bands to be included in the second variation. 103

6.1 Computation of $H\varphi$ with a non-local pseudopotential and the projector basis method. 117

Preface

The first edition of this book, published in 1994, provided an exposition of the LAPW method and its relationship with other electronic structure approaches, especially Car-Parrinello based planewave methods. Since publication of that book, the LAPW method has been transformed from a specialized method used mostly by researchers running their own home made versions, to a popular, widely used method, where most users run standard codes to investigate materials of interest to them. This is an exciting development because it opens the door to widespread use of first principles calculations in diverse areas of condensed matter physics and materials science. The positive impact of this on scientific progress is already becoming clear. Also as a result of this trend, the great majority of researchers using the LAPW method are no longer directly involved in the development of LAPW codes. Nonetheless, it remains important to understand how the LAPW method works, what its limitations are, and how its parameters determine the quality and efficiency of calculations. The scientist with an understanding of how the method works has a clear advantage. This edition is an updated and expanded treatment of the LAPW method, including descriptions of key developments in the LAPW method since 1994, such as $p_{1/2}$ local orbitals, the APW+LO method, LDA+U calculations and non-collinear magnetism, as well as much of the material from the first edition.

The exceptionally high accuracy and reasonable computational efficiency of the general potential linearized augmented planewave (LAPW) method has led to its emergence as the standard by which density functional calculations for transition metal and rare-earth containing materials are judged. Furthermore, the widespread availability of high quality, user-friendly LAPW codes has made it a very popular method for first principles studies of materials.

However, even though codes are generally provided in source form, it remains difficult for most users to delve into the method and implement calculations for properties of interest to them or simply to understand exactly how to set the parameters to optimally solve a given problem. Among the main obstacles is the fact the essential details about the LAPW method are scattered through the literature, and some crucial aspects such as how to set the linearization parameters and sphere radii are hardly discussed at all. Newcomers to the field have had to learn the LAPW method either from one of the groups that currently uses it, or by the arduous process of reconstructing the method after gathering and digesting the literature on it.

This book has two primary purposes. The first is addressed in the main part of this book, which is aimed at lowering this barrier so that newcomers can quickly learn the method and start performing calculations using an existing code, or, if desired, write a new code without having to "reinvent the wheel". An algorithmic approach is used to do this. Theory is discussed, but the emphasis is on how a practical implementation proceeds and on information that will be useful in carrying out calculations with LAPW codes. It is our hope that this edition will help new researchers more effectively use the LAPW method and its extensions to solve problems of particular interest to them, as well as advancing the method by providing a unified exposition of its inner workings.

Our second purpose in writing this book derives from the very rapid progress in planewave based electronic structure techniques since the development of the Car-Parrinello method. These and related approaches have greatly increased the range of systems that can be treated using planewave based methods, but the underlying ideas have been slow to be applied to other methods. It is our view that this will prove to be a very fruitful endeavor. Accordingly, we devote a portion of this book to discussion of the (1) relationships between the LAPW and planewave pseudopotential methods, (2) essential aspects of the Car-Parrinello method and (3) early work aimed at incorporating Car-Parrinello like algorithms into the LAPW method. We hope that this will both help readers understand the relationship between different electronic structure methods and stimulate further work in this area.

The ideas presented in this book regarding the relationships between LAPW and pseudopotential methods have been evolving for some time, and are the product not only of our thinking, but that of many others as well. We have benefited greatly from contributions of others and from many helpful discussions that we have had the opportunity to participate in. We especially thank Henry Krakauer, who taught one of us (DJS) the LAPW method and Warren E. Pickett who served as his mentor at the Naval Research Laboratory. A substantial part of this book is composed of things learned from them. Special thanks are also due to Karlheinz Schwarz and to Peter Blaha. Many of the insights in this book derive from ongoing interactions with them and from especially discussions in Vienna and at a series of workshops in Planneralm, Austria. Thanks are also due to R.E. Cohen, S. Goedecker, C. Haas, B.M. Klein, D.D. Koelling, A.Y. Liu, I.I. Mazin, V.L. Moruzzi, M. Posternak, E. Sjöstedt, S.H. Wei, E. Wimmer and R. Yu.

Most importantly, we thank our wives, Nancy and Hedvig, for their loving encouragement and support, without which this work would not have been possible. It is our pleasure to dedicate this book to them.

David J. Singh and Lars Nordström,

Chapter 1

INTRODUCTION

Over the past decades, our technological and industrial base has become increasingly dependent on advanced materials. There is every indication that this trend will continue to accelerate, and that progress in many areas will depend increasingly on the development of new materials and processing techniques.

A second and equally significant trend is the continuing ascent of the information technologies, which now touch almost every aspect of life in some way. Most significant for scientists is the emergence of numerical modeling as a powerful and widely used tool. This has been fueled not only by the increasing performance of state of the art supercomputers but also, and perhaps more importantly, by the availability of relatively inexpensive high performance workstations. In fact, the ubiquitous personal computer has become a capable research tool and all but the most demanding calculations can now be done routinely using widely available PC's and PC clusters. Problems that required the largest supercomputers in the early 1980's can now be solved routinely in almost any location.

In this environment, it is natural that there is a strong interest in using numerical modeling in materials science. Historically, progress in materials science has almost always occurred through laboratory experimentation, often guided by empirical trends, knowledge of the properties of related materials, and physical and chemical intuition, but less commonly by predictions based on numerical modeling. This is changing.

First principles simulations, using density functional theory and particularly the local density approximation and generalized gradient approximations, have proved to be a reliable and computationally tractable tool in condensed matter physics. These simulations have now impacted virtually every area of this broad field. Applications in materials science had, however, remained more elusive, since this is the realm not of the simple ordered stoichiometric solids that are most easily simulated, but of so called "real materials". In other words to apply first principles techniques in materials science it is necessary to treat complex systems, with stoichiometric deviations, surfaces, impurities, grain boundaries and other point and extended defects. However, we are now at the threshhold where simulations are starting to make a major impact in this area as well.

Along with the advances in computing technology that have occurred during the last decade, there have been important algorithmic improvements, particularly for planewave based methods, and for certain classes of materials it is now feasible to simulate systems containing several hundred atoms in a unit cell. This opens the door for the direct application of these techniques in studying a substantial set of real materials problems. Further, it is possible to use first principles calculations to create sophisticated data sets that can parameterize model Hamiltonians. These then can be used to model even more complex materials problems.

At this time, density functional practitioners are divided into two nearly disjoint communities; one employing pseudopotentials and relatively simple basis sets (particularly planewaves) and the other using methods with complex but efficient basis sets, such as the linearized augmented planewave (LAPW), the linearized muffin-tin orbital (LMTO) and related methods. The latter community has traditionally dominated research on transition metals and their compounds.

There are signs that this could change. The development of improved algorithms, particularly the Car-Parrinello (CP) and related methods, and of sophisticated ultrasoft pseudopotentials has for the first time made it feasible to perform accurate simulations of complex transition metal systems using planewave basis sets. Nonetheless, the LAPW, LMTO and related approaches remain the methods of choice for transition metals, and a case can be made that this will continue, especially since the algorithmic improvements, referred to generically as CP, can be incorporated into LAPW and other codes.

It is our view that a more realistic scenario is that these traditionally distinct approaches will eventually converge. Ultrasoft pseudopotentials bear relationships with the LAPW method, and pseudopotentials corresponding to the LAPW method have been explicitly constructed. It is perhaps not surprising that there are relationships between planewave pseudopotential methods and the LAPW method. Both approaches have a common starting point, *i.e.* a planewave basis set. Further, both approaches are motivated by the observation that planewaves, by themselves, are ill suited for direct solution of Schrodinger's equation in a crystal. This is because the potential and therefore the wavefunctions are rapidly varying near the nuclei. In the planewave pseudopotential approach, this problem is avoided by replacing the Hamiltonian near the atoms with a smoother pseudo-Hamiltonian in such a way that the valence energy spectrum is reproduced, but the core states are removed, as are the rapid variations in the wavefunctions near the nucleus. In the LAPW method, the planewaves are modified near the atoms rather than the Hamiltonian. This modification (the augmentation) is such that small wavevector (**G**) augmented planewaves can reproduce the rapid variations in the valence wavefunctions. In addition the augmented planewaves are made orthogonal to the core states. Thus, as in the pseudopotential method, the valence energy spectrum is reproduced with a low planewave cutoff, and the core states are removed from the spectrum. In both cases, the basis functions are labeled as planewaves (by **G**), the Hamiltonian matrix elements are modified in such a way that rapid convergence with the maximum $|\mathbf{G}|$ is obtained, and the modification due to an atom is in the vicinity of that atom.

Viewed as a pseudopotential method, the LAPW method is the ultimate in ultrasoft pseudopotentials; extremely well converged calculations for $3d$ transition metals and even f-electron systems require plane-wave cut-offs ranging from 10 to 20 Rydbergs. Further, as mentioned, algorithmic improvements that had been confined to planewave basis sets are now beginning to be exploited in the LAPW and other methods. In the final chapter of this book, the all-electron LAPW method will be reformulated to look like a planewave pseudopotential approach.

The purpose of this monograph is two-fold. First of all, it provides a detailed and self-contained exposition of the LAPW method, so that newcomers to the field can quickly learn the technique, and if desired construct a working code. Secondly, connections are made between the LAPW and planewave pseudopotential points of view. We hope that this material will make the LAPW method more understandable to readers experienced with planewave methods, and will help stimulate work at the interface of these traditionally distinct approaches.

The organization of this book follows these two themes. Readers seeking an exposition of the LAPW method will find it in Chapters 4 and 5; these are self-contained. The remaining chapters are devoted to building bridges between the LAPW method and planewave Car-Parrinello approaches, and a reading of these plus Chapter 4, would be a good strategy for readers interested particularly in this topic.

Chapter 2

DENSITY FUNCTIONAL THEORY AND METHODS

Condensed matter physics and materials science are concerned fundamentally with understanding and exploiting the properties of interacting electrons and atomic nuclei. This has been well known since the development of quantum mechanics. With this comes the recognition that, at least in principal, almost all properties of materials can be addressed given suitable computational tools for solving this particular problem in quantum mechanics. Unfortunately, the electrons and nuclei that compose materials comprise a strongly interacting many body system, and this makes the direct solution of Schrodinger's equation an extremely impractical proposition. Rather, as was stated concisely by Dirac in 1929, progress depends on the development of sufficiently accurate, but tractable, approximate techniques [41].

Thus the development of density functional theory (DFT) and the demonstration of the tractability and accuracy of the local density approximation (LDA) to it defined an important milestone in condensed matter physics. First principles quantum mechanical calculations based on the LDA and extensions, like generalized gradient approximations, have emerged as one of the most important components of the theorist's toolbox. These methods are also starting to have significant impact in many areas of materials science, though there remains much to be done. A real challenge is posed by the highly complex nature of most real materials. Related to this, there has been considerable progress in developing DFT based methods suitable for large systems containing many hundreds of atoms in a unit cell. In addition, widely available user friendly DFT codes, with implementation of many property calculations are available. It seems very reasonable to expect these trends to continue and for DFT calculations to become ubiquitous tools in materials science.

It is worth noting that the DFT of Hohenberg and Kohn [73] was predated by the LDA, which was developed and applied by Slater [177] and his co-workers (see Ref. [179]). Nonetheless, the impact of local density approximation (LDA) calculations in solid state physics remained limited until the late 1970's, when several calculations demonstrating the feasibility and accuracy of the approach in determining properties of solids appeared [215, 216, 217, 120]. The contribution of these and other pioneers in this field should not be underestimated. There has been a great deal written about why the LDA should or should not be adequate for calculating properties of this or that material. There is, however, no doubt that the most convincing arguments derive from the direct comparison of detailed calculations with experiment. The utility of the LDA was demonstrated in the early calculations of these and other groups, and it is this that has led to the widespread application of these tools.

As mentioned, DFT based calculations have become one of the most frequently used theoretical tools in condensed matter physics, and there are now several excellent reviews of the subject including those by Lundqvist and March, [112] Callaway and March, [29], Dreizler and da Provincia, [42] Ernzerhof, Perdew and Burke [44] and Parr and Yang [137]. The reader is warned that this chapter is not along those lines, *i.e.* it is not a comprehensive review of DFT. Rather its purpose is much more limited – to present a very limited sketch, emphasizing those aspects that are necessary groundwork for the material to follow. For a general exposition of DFT, the reader is referred to the excellent reviews mentioned above.

2.1 Density Functional Theory

The theorem upon which DFT and the LDA are based is that of Hohenberg and Kohn. It states that the total energy, E, of a non-spin-polarized system of interacting electrons in an external potential (for our purposes the Coulomb potential due to the nuclei in a solid) is given exactly as a functional of the ground state electronic density, ρ.

$$E = E[\rho]. \tag{2.1}$$

They further showed that the true ground state density is the density that minimizes $E[\rho]$, and that the other ground state properties are also functionals of the ground state density. The extension to spin-polarized systems is straightforward; E and the other ground state properties become functionals of the spin density, which in the general case is given as a four component spinor [198, 155]. In the collinear case, where the spin-up and spin-down densities suffice,

$$E = E[\rho_\uparrow, \rho_\downarrow]. \tag{2.2}$$

Unfortunately, the Hohenberg-Kohn theorem provides no guidance as to the form of $E[\rho]$, and therefore the utility of DFT depends on the discovery of sufficiently accurate approximations. In order to do this, the unknown functional, $E[\rho]$, is rewritten as the Hartree total energy plus another, but presumably smaller, unknown functional, called the exchange-correlation (xc) functional, $E_{xc}[\rho]$.

$$E[\rho] = T_s[\rho] + E_{ei}[\rho] + E_H[\rho] + E_{ii}[\rho] + E_{xc}[\rho]. \tag{2.3}$$

Here $T_s[\rho]$ denotes the single particle kinetic energy, $E_{ei}[\rho]$ is the Coulomb interaction energy between the electrons and the nuclei, $E_{ii}[\rho]$ arises from the interaction of the nuclei with each other, and $E_H[\rho]$ is Hartree component of the electron-electron energy,

$$E_H[\rho] = \frac{e^2}{2}\int \mathrm{d}^3\mathbf{r}\mathrm{d}^3\mathbf{r}' \,\frac{\rho(\mathbf{r})\rho(\mathbf{r}')}{|\mathbf{r}-\mathbf{r}'|}. \tag{2.4}$$

As mentioned, $E_{xc}[\rho]$ is an unknown functional. However, several useful approximations to it are known. The simplest is the local density approximation (LDA). In the LDA, $E_{xc}[\rho]$ is written as

$$E_{xc}[\rho] = \int \mathrm{d}^3\mathbf{r}\;\; \rho(\mathbf{r})\epsilon_{xc}(\rho(\mathbf{r})), \tag{2.5}$$

where $\epsilon_{xc}(\rho)$ is approximated by a local function of the density, usually that which reproduces the known energy of the uniform electron gas. The other commonly used approximations are the generalized gradient approximations (GGAs), where the local gradient as well as the density is used in order to incorporate more information about the electron gas in question, *i.e.* $\epsilon_{xc}(\rho)$ is replaced by $\epsilon_{xc}(\rho, |\nabla\rho|)$. The weighted density approximation (WDA) is an approximation that incorporates more non-local information about the electron gas via a model pair correlation function. This is exact in important limits: the uniform electron gas and arbitrary single electron systems. It greatly improves the energies of atoms, and often yields bulk properties that are much improved as well. Nonetheless, the WDA is more computationally demanding than the LDA or GGA, and as such relatively few WDA studies have been reported for solids.

The relationship between the various approximations can be understood using the exact expression for the exchange correlation energy in terms of the pair correlation function [100, 143],

$$E_{xc}[n] = \int\int \mathrm{d}^3\mathbf{r}\mathrm{d}^3\mathbf{r}' \;\frac{n(\mathbf{r})n(\mathbf{r}')}{|\mathbf{r}-\mathbf{r}'|}\bar{g}[n,\mathbf{r},\mathbf{r}'] \tag{2.6}$$

$$= \int\int \mathrm{d}^3\mathbf{r}\mathrm{d}^3\mathbf{r}' \frac{n(\mathbf{r})\bar{n}_{xc}[n,\mathbf{r},\mathbf{r}']}{|\mathbf{r}-\mathbf{r}'|}, \tag{2.7}$$

where $\bar{g}$ is the coupling constant average (from e^2=0 to e^2=1 in atomic units) of the pair correlation function of the electron gas in question, and $\bar{n}_{xc}$ is the coupling constant averaged exchange correlation hole. Since the exchange correlation hole must be a depletion containing exactly one electron charge, E_{xc} is invariably negative. The physical meaning of this expression is that the exchange correlation energy is given by the Coulomb interaction of each electron with its exchange correlation hole, reduced in magnitude by a kinetic energy contribution, which corresponds to the energy required to dig out the hole. This reduction is accounted for by using the coupling constant average instead of the full strength pair correlation function, and includes contributions to the kinetic energy beyond the single particle level. The spherically symmetric Coulomb interaction in Eqn. 2.7 means that only the spherical average of the exchange correlation hole needs to be correct to obtain the correct energy, a fact that is important in the success of simple approximations like the LDA [112, 44].

The local density approximation consists of the replacement in Eqn. 2.7,

$$n(\mathbf{r}')\bar{g}[n,\mathbf{r},\mathbf{r}'] \rightarrow n(\mathbf{r})\bar{g}^h(n(\mathbf{r}),|\mathbf{r}-\mathbf{r}'|), \tag{2.8}$$

where $\bar{g}^h$ is the function $\bar{g}$ for the uniform electron gas. This reproduces the exact energy for the uniform electron gas. The weighted density approximation retains the non-locality using integration with a model function $\bar{g}^w$.

$$n(\mathbf{r}')\bar{g}[n,\mathbf{r},\mathbf{r}'] \rightarrow n(\mathbf{r})'\bar{g}^w(\bar{n}(\mathbf{r}),|\mathbf{r}-\mathbf{r}'|), \tag{2.9}$$

where $\bar{n}(\mathbf{r})$ is the weighted density, determined using the sum rule,

$$\int \mathrm{d}^3\mathbf{r}'n(\mathbf{r}')[\bar{g}^w(\bar{n}(\mathbf{r}),|\mathbf{r}-\mathbf{r}'|)-1] = -1. \tag{2.10}$$

This approximation violates the exact symmetry $\bar{g}[n,\mathbf{r},\mathbf{r}'] = \bar{g}[n,\mathbf{r}',\mathbf{r}]$, but nonetheless has been quite successful in describing structural properties of materials in the admittedly few calculations reported to date.

Modern GGA functionals appear formally like the LDA, but with the local function including not just the density (or spin densities in the LSDA), but also the local gradient (or spin density gradients). However, these are not gradient expansions, but rather sophisticated methods to obtain as good an energy as possible using known exact sum rules and scaling relationships for the electron gas based on Eqn. 2.7 and/or fits to data bases [102, 141, 142, 44, 11, 12,

13, 14]. In contrast to the WDA, GGA calculations have been performed for a wide variety of materials, and the GGA is in fact the method of choice for many first principles studies of materials. The behavior of the GGA relative to the LDA is well understood on the basis of the many comparative studies that have been done. From the results of these, the following conclusions may be drawn: (1) GGAs significantly improve the ground state properties of light atoms and molecules, clusters and solids composed of them; (2) many properties of $3d$ transition metals are greatly improved; for example, unlike the LSDA the correct *bcc* ground state of Fe is obtained; (3) the description of Mott-Hubbard insulators, like the undoped phases of high-T_c cuprates, is not significantly improved over the LSDA; (4) GGA functionals usually favor magnetism more than the LSDA, and as a result the magnetic energies for some $3d$ transition metals may be overestimated; and (5) structural properties are generally improved, although GGAs sometimes lead to overcorrection of the LDA errors in lattice parameters. In some materials containing heavy elements (*e.g.* in $5d$ compounds) these degrade agreement with experiment relative to the LDA. It should be emphasized, however, that there is ongoing work aimed at developing even better GGA functionals, and it is quite possible that an improved form that alleviates the above deficiencies will be found.

Kohn and Sham [90] wrote the electron density as a sum of single particle densities, and used the variational property to obtain a prescription for determining the ground state energy and density, given the functional E_{xc}. In particular, they showed that the correct density is given by the self-consistent solution of a set of single particle Schrodinger-like equations, known as the Kohn-Sham (KS) equations, with a density dependent potential,

$$\{T + V_{ei}(\mathbf{r}) + V_H(\mathbf{r}) + V_{xc}(\mathbf{r})\}\, \varphi_i(\mathbf{r}) = \epsilon_i \varphi_i(\mathbf{r}), \tag{2.11}$$

with the density given by a Fermi sum over the occupied orbitals,

$$\rho(\mathbf{r}) = \sum_{occ} \varphi_i^*(\mathbf{r})\varphi_i(\mathbf{r}). \tag{2.12}$$

Here the highest occupied orbital is determined by the electron count, the φ_i are the single particle Kohn-Sham orbitals, the ϵ_i are the corresponding Kohn-Sham eigenvalues, T is the single particle kinetic energy operator, V_{ei} is the Coulomb potential due to the nuclei, V_H is the Hartree potential and V_{xc} is the exchange correlation potential. Both V_H and V_{xc} depend on ρ.

$$V_H(\mathbf{r}) = e^2 \int \mathrm{d}^3\mathbf{r}' \frac{\rho(\mathbf{r}')}{|\mathbf{r} - \mathbf{r}'|}, \tag{2.13}$$

and

$$V_{xc}(\mathbf{r}) = \frac{\delta E_{xc}[\rho]}{\delta \rho(\mathbf{r})}. \tag{2.14}$$

In this framework, a calculation entails the self-consistent solution of Eqns. 2.11 and 2.12. That is, a density must be found such that it yields an effective potential that when inserted into the Schrodinger-like equations yields orbitals that reproduce it. Thus, instead of having to solve a many-body Schrodinger equation, using DFT we have the far easier problem of determining the solution to a series of single particle equations, along with a self-consistency requirement.

In solids, Bloch's theorem provides a further simplification that facilitates DFT based calculations: Because the charge density has the periodicity of the lattice, so does the single particle KS Hamiltonian. Thus KS orbitals with different Bloch momenta are coupled only indirectly through the density dependent potential. Accordingly, in DFT based (but not, for example, in Hartree-Fock) calculations, the single particle KS equations may be solved separately on a grid of sampling points in the symmetry irreducible wedge of the Brillouin zone, and the resulting orbitals used to construct the charge density.

2.2 Solution of the Single Particle Kohn-Sham Equations

DFT based electronic structure methods are classified according to the representations that are used for the density, potential and, most importantly, the KS orbitals. The choice of representation is made to minimize the computational and human (*e.g.* programming) costs of calculations, while maintaining sufficient accuracy. These competing material and application dependent goals have led to the development and use of a wide variety of techniques.

This book is concerned with two particular approaches, planewave pseudopotential methods and the LAPW method. It is certainly possible to avoid the explicit use of a basis in constructing the KS orbitals, for example, by numerically solving the differential equations on grids. However, nearly all approaches that have been proposed for solids, including the planewave pseudopotential and LAPW methods, do rely on a basis set expansion for the KS orbitals. Here the discussion is confined to methods that do use a basis, in which case the KS orbitals are:

$$\varphi_i(\mathbf{r}) = \sum_{\alpha} c_{i\alpha} \phi_\alpha(\mathbf{r}), \tag{2.15}$$

where the $\phi_\alpha(\mathbf{r})$ are the basis functions and the $c_{i\alpha}$ are expansion coefficients. Since, given a choice of basis, these coefficients are the only variables in the problem (note that the density depends only on the KS orbitals), and since the total energy in DFT is variational, solution of the self-consistent KS equations amounts to determining the $c_{i\alpha}$ for the occupied orbitals that minimize the total energy.

To proceed, note that the energy can be rewritten using the single particle eigenvalues to eliminate the unknown functional, $T_s[\rho]$.

$$E[\rho] = E_{ii}[\rho] + \sum_{occ} \epsilon_i + E_{xc}[\rho] - \int \mathrm{d}^3\mathbf{r}\rho(\mathbf{r})(V_{xc}(\mathbf{r}) + \frac{1}{2}V_H(\mathbf{r})), \quad (2.16)$$

where the sum is over the occupied orbitals, and ρ, V_H and V_{xc} are given by Eqns. 2.12, 2.13 and 2.14, respectively.

It is very common to separate the determination of the $c_{i\alpha}$ and the determination of the self-consistent charge density in density functional calculations. The solutions for the density and the $c_{i\alpha}$ are done hierarchically in this case, as shown schematically in Fig. 2.1. In this scheme, it is necessary to repeatedly determine the $c_{i\alpha}$ that solve the single particle equations 2.11 for fixed charge density. This may be done using standard matrix techniques. Specifically, given the basis, the Kohn-Sham Hamiltonian and overlap matrices, $\mathbf{H}$ and $\mathbf{S}$ are constructed and the matrix eigenvalue equation,

$$(\mathbf{H} - \epsilon_i\mathbf{S})\mathbf{c}_i = 0, \quad (2.17)$$

is solved at each **k**-point in the irreducible wedge of the Brillouin zone. This can be done efficiently using standard linear algebra routines, such as EISPACK. Here the square matrices $\mathbf{H}$ and $\mathbf{S}$ are of rank equal to the number of basis functions, n_b and the $\mathbf{c}_i$ are vectors containing the n_b coefficients, $c_{i\alpha}$ for each KS orbital i.

If the true occupied KS orbitals can be expressed as linear combinations of the basis functions, then optimizing the $c_{i\alpha}$ will yield the exact self-consistent solution. On the other hand, if the exact KS orbitals cannot be expressed exactly in terms of the chosen basis, this procedure will yield an approximate solution that is optimal in the sense that it gives the lowest possible total energy for this basis. The quality of a basis set can, therefore, be measured by the extent to which the total energy evaluated using the orbitals of Eqn. 2.15 differs from the true KS energy.

Efficiency, bias, simplicity and completeness are common terms that are used in discussing the relative merits of electronic structure techniques. These refer to the number of the basis functions needed to achieve a given level of convergence, whether or not the basis favors certain regions of space over others (*e.g.* by being more flexible near atomic nuclei than elsewhere), the difficulty in calculating matrix elements and whether the basis can be improved arbitrarily by adding additional functions of the same type.

Planewave basis sets are notoriously inefficient in the above sense for most solids. This, however, is not necessarily a defect, since it just reflects the fact that they are unbiased. Further, planewaves form a complete set and they are a simple basis. The completeness means that, at least in principle, arbitrary accuracy can be obtained by increasing the number of planewaves in the basis, and more importantly that the convergence of a calculation can be monitored by varying the planewave cutoff. Further, because of the simplicity of this basis, implementation of planewave codes is relatively straightforward, and matrix elements of many operators can be calculated quickly. The fact that wavefunctions expanded in planewaves can be transformed efficiently from reciprocal space (coefficients of the planewave expansion) to real space (values on a real space grid) using fast Fourier transforms (FFTs) means that many operators can be made diagonal. In particular, the kinetic energy and momentum operators are diagonal in reciprocal space, and the operation of local potentials is diagonal in real space.

It is apparent from Eqn. 2.15 that the most efficient basis set consists of the KS orbitals themselves (or equivalently, linear combinations of the KS orbitals). In this case, an exact calculation is achieved using a basis set size equal to the number of occupied orbitals. Even though, in general, the KS orbitals are unknown at the beginning of a calculation, this property can be exploited in constructing basis sets. In particular, if the KS orbitals for a similar Hamiltonian are known, inclusion of these in the basis will often result in a great improvement. A simple example is the case where a small perturbation, ΔH (*e.g.* spin-orbit) is added to a Hamiltonian, H_0 for which a solution has already been generated. Using the KS orbitals of H_0 as a basis, the matrix elements with the perturbed Hamiltonian can be readily constructed as those of ΔH with the addition of the eigenvalues of H_0 on the diagonal. The construction and diagonalization of the Hamiltonian in this space can often be done quite rapidly, even for complicated ΔH because of its small dimension.

More common examples are the use of atomic and muffin-tin orbitals in electronic structure calculations. In the former, an atomic H_0 is assumed in constructing basis functions for each site. Despite the fact that crystal potentials are often significantly different from atomic potentials, even in the vicinity of an ion, linear combination of atomic orbitals (LCAO) methods have been quite successful, particularly for large systems, where the efficiency of this basis is an important advantage. However, although this basis is clearly complete, problems often arise when attempts are made to add large numbers of basis functions to obtain highly converged calculations. This is because atomic orbitals centered at a single site are already complete. Thus LCAO's which have orbitals centered at each site are over-complete and, because of this, the overlap matrix, $\mathbf{S}$, in Eqn. 2.17 becomes ill-conditioned for large basis sets.

Muffin-tin orbital derived basis sets will be discussed in the chapters on the LAPW method. Here it suffices to state that they are based on solutions of radial Schrodinger's equation with a better approximation to the crystal potential in the vicinity of the site in question than that used in constructing LCAOs and that methods using them can be constructed to largely avoid the over-completeness problem.

2.3 Self-Consistency in Density Functional Calculations

As mentioned, the Hohenberg-Kohn theorem shows that the total energy is variational, and this is the key to its usefulness. The true ground state density is that density which minimizes the energy. When approximations are made to $E_{xc}[\rho]$, such as the LDA, there is no longer a true variational principle, and there is no guarantee that the energy obtained by minimizing the now approximate energy functional will be higher than the exact ground state energy. Clearly then, the relative quality of different approximations cannot be determined by determining which of them yields the lower energy. Furthermore, the true ground state density is not in general the density that minimizes the total energy as determined using approximate functionals. There is, in fact, no prescription for determining what the exact ground state density is from approximate functionals. Accordingly, calculations proceed by minimizing the approximate energy functional, recognizing that, although the resulting energy may be lower (or higher) than the true ground state energy, a good approximation to the energy functional should give a good energy and density and that the procedure is exact for the true energy functional.

Since the single particle kinetic energy, $T_s[\rho]$ appearing in Eqn. 2.3 is unknown in this form, the minimization proceeds via the KS equations. Then the variation is with respect to the orbitals, or in a basis set expansion, the coefficients, $c_{i\alpha}$. With a fixed basis, these are the only parameters that can be varied; otherwise there are additional parameters that determine the basis functions. In any case, the problem may be stated as follows: Find the coefficients (and other parameters, if any) that minimize the energy functional, Eqn. 2.16, subject to the constraint that the orbitals remain orthonormal.

The direct minimization of the total energy with respect to the $c_{i\alpha}$ was proposed quite early on by Bendt and Zunger [15] (see also Payne *et al.* [139]) and is at the heart of the Car-Parrinello (CP) and related methods [30]. Nonetheless, and in spite of potential computational advantages, this type of approach has not yet become popular for methods that use non-planewave basis sets. This is a result of the complexity of the optimization problem; there are typically hundreds or thousands of parameters even for small problems and the objective function is highly nonlinear with many local extrema (corresponding to missed KS orbitals, with occupation of higher lying ones).

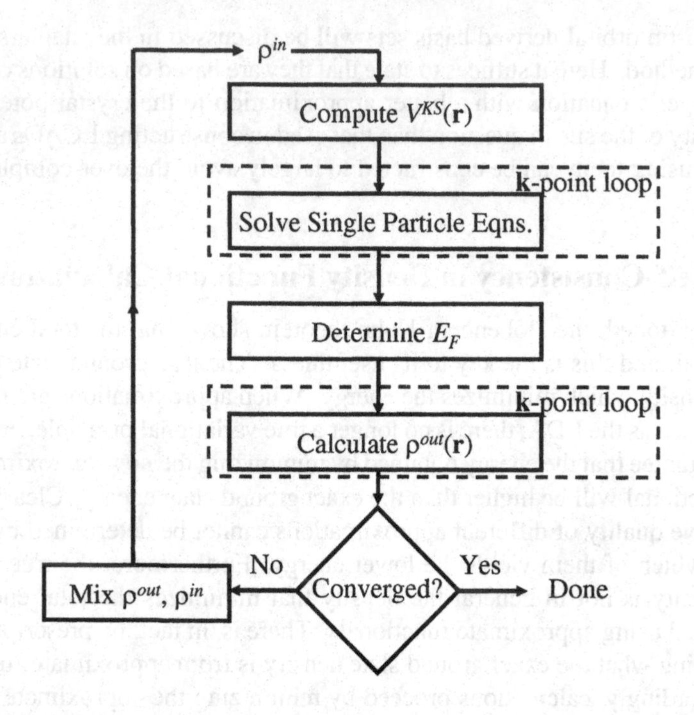

Figure 2.1. Schematic fbw-chart for self consistent density functional calculations.

Because of these complications, the historically dominant approach has been to refine the density iteratively by solving Eqns. 2.11 and 2.12 alternately. This is the basis of the standard self-consistency cycle illustrated in Fig. 2.1. Given a charge density, Eqn. 2.17 is diagonalized, ensuring that the orbitals are orthonormal and that no orbitals are missed. This eliminates almost all local extrema. An output charge density is constructed from the eigenvectors using Eqn. 2.12, and then mixed with the input to yield a refined input for the next iteration. The simplest mixing scheme is straight mixing:

$$\rho_{in}^{i+1} = (1-\alpha)\rho_{in}^{i} + \alpha\rho_{out}^{i}, \tag{2.18}$$

where the superscript refers to the iteration number and α is the mixing parameter. For sufficiently small α, the iterations converge. However, the radius of convergence can be small, and rapidly becomes smaller as the size of the unit cell increases, particularly for metallic and/or magnetic systems. As a result, considerable effort has been devoted to devising more sophisticated mixing procedures, using information from previous iterations to accelerate the convergence. The most common of these is Broyden's method and variants thereof [28, 186]. These will be discussed in detail in Chapter 5.

As mentioned, this is a hierarchical approach to the optimization. The diagonalization may viewed as an optimization (minimization of the residuals); this is the lowest level of the hierarchy. The next level, which can also be regarded as an optimization (minimization of the difference between the input and output densities) is the search for a self-consistent charge density. Experience has shown that this approach is very robust. It is, however, inefficient in the following sense.

Exact eigenvectors are calculated (Eqn. 2.17) for the current single particle Hamiltonian at each step of the iteration to self-consistency, including early iterations for which the charge density is poor. However, these are of little interest; the only eigenvectors that are relevant are those for the self-consistent charge density. In the early iterations, approximate eigenvectors would serve as well. This observation provides useful a insight into the CP method, and suggests avenues for speeding up non-planewave approaches.

The eigenvectors and the input charge density may be viewed as independent quantities to be optimized (recall that at the minimum the input and output densities are equal removing this independence). The iteration to self-consistency is nothing more than a sequence of moves towards the minimum. In the hierarchical approach, the eigenvector moves are to the exact solution for the current density, while the moves of the charge density are determined by the mixing. Given the true charge density, a single move of the eigenvectors yields the true minimum. Meanwhile, the complex nonlinear dependence of the single particle Hamiltonian on the density makes the charge density moves less effective.

In the CP method, the eigenvector moves are based on an iterative refinement rather than exact diagonalizations. One or more steps of an iterative diagonalization are used to generate the refined eigenvectors, which are then used to construct a charge density move. For planewave basis sets, FFT dependent algorithms (discussed in Chapter 3) can be used to perform these refinements in a small fraction of the time needed for exact diagonalizations. Thus, even though there may be an increase in the total number of iterations needed, their cost is much lower, particularly for large systems. This underlies, at least in part, the efficiency of the CP method.

The question often arises as to why CP like algorithms have not yet been widely applied in non-planewave methods. Certainly, the basic idea of iteratively refining the eigenvectors along with the charge density is applicable to any method using a basis set. The first complication is that, in order for CP based algorithms to be worthwhile, an eigenvector move based on iterative refinement should be much faster than one using exact diagonalization. The second is that these refinements need to be effective in the sense of rapidly converging to the true eigenvectors. This can be difficult in techniques with non-orthogonal basis sets and poorly conditioned matrices, including the LAPW method.

As mentioned, the main raison d'etre of non-planewave basis sets is to reduce the size of the secular equation (Eqn. 2.17) for materials with hard pseudopotentials. In general, the price to be paid for this efficiency is an increase in the cost of computing matrix elements. In planewave based methods, the cost of synthesizing the Hamiltonian matrix is often negligible compared to the cost of diagonalizing it. However, it is often the case with non-planewave methods that the cost of synthesizing the Hamiltonian matrix rivals or even exceeds the cost of diagonalizing it.

The key step in an iterative refinement is the operation of the Hamiltonian on wavefunctions. This can be done either by synthesizing the Hamiltonian matrix and then performing matrix-vector multiplies, or directly as in the CP method.

In the LAPW method, as it is normally implemented, the cost of computing the Hamiltonian and overlap matrices is smaller than the diagonalization time, but only by a factor of two to five, depending on the details of the system (it scales with system size in the same way as the diagonalization). This limits the gains that can be obtained by adopting an iterative refinement of the eigenvectors if the Hamiltonian matrix is synthesized. On the other hand, the cost of operating the Hamiltonian directly on a wavefunction is at least equal to the cost of computing a single row of the Hamiltonian matrix. Further, in the LAPW method the dimension of the Hamiltonian is typically only an order of magnitude larger than the number of occupied states. Since the number of operations of the Hamiltonian in iterative approaches is several times the number of orbitals (depending on the exact scheme used), the potential gains from this approach are limited as well. What is needed then are new algorithms for operating the Hamiltonian on trial wavefunctions, *i.e.* linear combinations of the basis functions. Progress for the LAPW method in this direction is discussed in the final chapter of this book.

2.4 Spin-Polarized Systems

In the generalization of DFT to spin-polarized systems [198], the charge density is augmented by a magnetization density, $\mathbf{m}(\mathbf{r})$. This is in general a continuous three dimensional vector field; both the magnitude and direction of $\mathbf{m}(\mathbf{r})$ vary from place to place. In nature, magnetism is often non-collinear, *i.e.* the magnetization direction does in fact vary from place to place. This non-collinearity arises for many reasons [156, 157], *e.g.* Fermi surface effects leading to spin spirals, frustration of exchange interactions as in triangular lattice systems, or between spin-orbit and exchange, like in U_3P_4, and other relativistic effects like the Dzyaloshinskii-Moriya interaction, which leads to the helical magnetic order of MnSi. However, many interesting magnetic systems either are collinear or are well approximated as collinear. In this case, which we discuss first, the direction dependence of $\mathbf{m}(\mathbf{r})$ reduces to a sign and therefore the theory may be formulated in terms of two scalar fields, a spin-up density, $\rho_{\uparrow}(\mathbf{r})$ and a spin-down density, $\rho_{\downarrow}(\mathbf{r})$. Then

$$\rho(\mathbf{r}) = \rho_{\uparrow}(\mathbf{r}) + \rho_{\downarrow}(\mathbf{r}), \tag{2.19}$$

and

$$m(\mathbf{r}) = \rho_{\uparrow}(\mathbf{r}) - \rho_{\downarrow}(\mathbf{r}). \tag{2.20}$$

In this case, the Hohenberg-Kohn theorem is generalized to state that the true ground state total energy is a variational functional of the spin densities [198, 45].

$$E = E[\rho, \mathbf{m}] = E[\rho_{\uparrow}, \rho_{\downarrow}], \tag{2.21}$$

where the first part of the equation applies in the general non-collinear case as well. The energy may then be decomposed as in Eqn. 2.3. The Coulomb terms remain functionals of the total density, but T_s and E_{xc} become functionals of the two spin densities. The variational principle is invoked to generate the spin-polarized KS equations of spin density functional theory.

$$(T + V_{ei}(\mathbf{r}) + V_H(\mathbf{r}) + V_{xc,\sigma}(\mathbf{r}))\varphi_{i\sigma}(\mathbf{r}) = \epsilon_{i\sigma}\varphi_{i\sigma}(\mathbf{r}), \tag{2.22}$$

where σ is the spin index and

$$\rho_{\sigma}(\mathbf{r}) = \sum_{occ} \varphi^*_{i\sigma}(\mathbf{r})\varphi_{i\sigma}(\mathbf{r}), \tag{2.23}$$

with the highest occupied orbital again determined by the electron count and

$$V_{xc,\sigma} = \frac{\delta E_{xc}[\rho_{\uparrow}, \rho_{\downarrow}]}{\delta \rho_{\sigma}(\mathbf{r})}. \tag{2.24}$$

The total energy expression then becomes

$$\begin{aligned} E &= E_{ii} + \sum_{occ} \epsilon + E_{xc}[\rho_{\uparrow}, \rho_{\downarrow}] - \frac{1}{2}\int \mathrm{d}^3\mathbf{r} V_H(\mathbf{r})\rho(\mathbf{r}) - \\ &\int \mathrm{d}^3\mathbf{r} \left\{\rho_{\uparrow}(\mathbf{r})V_{xc,\uparrow}(\mathbf{r}) + \rho_{\downarrow}(\mathbf{r})V_{xc,\downarrow}(\mathbf{r})\right\}, \end{aligned} \tag{2.25}$$

where we are implicitly using the fact that the Hartree potential of a Coulomb system is twice the Hartree energy.

These equations are to be solved self-consistently, as in the non-spin-polarized case. The differences are:

1 The density is replaced by two spin densities.

2 There are separate sets of KS orbitals for the two spin components, and two sets of single particle equations need to be solved to obtain them.

3 V_{xc} is spin dependent; this is the only term in the single particle Hamiltonian that is explicitly spin-dependent.

4 In the total energy expression E_{xc} is a functional of the two spin densities. E_{xc} favors spin polarized solutions, T_s opposes them. Whether or not a material is magnetic depends on the balance between these terms.

Finally, because of the additional degrees of freedom contained in the spin density, spin-polarized KS equations often have multiple self-consistent solutions, corresponding to different stable spin configurations. Determining which of these is the ground (lowest energy) state and if there are any solutions that have been missed may require an exhaustive search. However, a constrained density functional technique, known as the fixed spin-moment method [205, 158], greatly simplifies the search in ferromagnetic systems. This procedure is discussed in chapter 5.

2.5 Non-Collinear Magnetism

In the case of general non-collinear magnetizations, the spin-dependent KS equations will no longer decouple. Instead it is meaningful to introduce an exchange-correlation field, which plays the role of an internal magnetic field, $\mathbf{b}_{\mathrm{xc}}$ [156, 175]. This is given by the functional derivative of the exchange-correlation energy with respect to the magnetization

$$\mathbf{b}_{\mathrm{xc}}(\mathbf{r}) = -\frac{\delta E_{xc}[\rho, \mathbf{m}]}{\delta \mathbf{m}(\mathbf{r})} . \tag{2.26}$$

With this auxiliary field, we can write the KS equation

$$\left(T + V_{ei}(\mathbf{r}) + V_H(\mathbf{r}) + V_{xc}(\mathbf{r}) - \mathbf{b}_{xc} \cdot \boldsymbol{\sigma}\right) \varphi_i(\mathbf{r}) = \epsilon_i \varphi_i(\mathbf{r}) , \tag{2.27}$$

which is now in a spinor form, *i.e.* $\varphi_i(\mathbf{r})$ is a general two-component spinor

$$\varphi_i(\mathbf{r}) = \begin{pmatrix} \alpha_i(\mathbf{r}) \\ \beta_i(\mathbf{r}) \end{pmatrix} . \tag{2.28}$$

In Eqn. 2.27, the first four operators are spin-independent, *i.e.* they act in the same way on both spin components, while the last operator is spin dependent through the action of the Pauli spin matrix vector, with its Cartesian components

$$\sigma_x = \begin{pmatrix} 0 & 1 \\ 1 & 0 \end{pmatrix}, \; \sigma_y = \begin{pmatrix} 0 & -i \\ i & 0 \end{pmatrix}, \; \sigma_z = \begin{pmatrix} 1 & 0 \\ 0 & -1 \end{pmatrix}. \quad (2.29)$$

The spin-independent part of the xc potential, $V_{xc}(\mathbf{r})$, is here the functional derivative of E_{xc} with respect to the charge density. It is now easy to see, by comparing Eqns. 2.22 and 2.27, that whenever the x and y components of the xc magnetic field vanish we return to the collinear case with decoupled equations for the spin components. Then $V_{xc,\sigma} = V_{xc} - \sigma b_{xc,z}$, where σ here is $\pm$ depending on spin character.

In order to get an iterative cycle, we need to obtain the charge and magnetization densities from our eigen-spinors in 2.28. This is done via

$$\rho(\mathbf{r}) = \sum_{\text{occ}} \varphi_i^\dagger(\mathbf{r})\varphi_i(\mathbf{r}) \quad (2.30)$$

and

$$\mathbf{m}(\mathbf{r}) = \sum_{\text{occ}} \varphi_i^\dagger(\mathbf{r})\boldsymbol{\sigma}\varphi_i(\mathbf{r}). \quad (2.31)$$

This scheme [132], which does not build in any geometrical approximations for the charge, magnetization, potential, nor xc magnetic field, *i.e.* a full potential formulation, has now been implemented in various electronic structure codes.

In the LSDA [198], the exchange-correlation energy is given by Eq. 2.5, but with a spin polarized energy density $\epsilon_{xc}(\rho_\uparrow(\mathbf{r}), \rho_\downarrow(\mathbf{r}))$. This gives a xc magnetic field from Eq. 2.26 that is locally parallel to the magnetization at each point in space

$$\mathbf{b}_{\text{xc}}(\mathbf{r}) = -\widehat{\mathbf{m}}(\mathbf{r})\,\rho(\mathbf{r}) \left[\frac{\partial \varepsilon_{xc}(\rho_\uparrow, \rho_\downarrow)}{\partial m}\right]_{\rho=\rho(\mathbf{r}), m=|\mathbf{m}(\mathbf{r})|}, \quad (2.32)$$

where $\widehat{\mathbf{m}}(\mathbf{r})$ is the unit vector along the direction of the magnetization density at point $\mathbf{r}$. For other approximations to the exchange correlation energy, the magnetic field is not as simple, and does not necessarily point in the same direction as the magnetization. Unfortunately, a complete GGA formulation has not yet been developed. In particular, all present formulations neglect gradients of the transverse component of the magnetization density. However, these gradients might not be essential, and there have been applications with these incomplete GGA formulae.

The secular matrix problem, Eq. 2.17, that enters a non-collinear calculation has double the size of that in a non-magnetic calculation or equivalently double the size of the secular equation for each of the individual spin components in a collinear calculation. That said, it is clear that these calculations are more time consuming. In addition, the relevant magnetic cell is usually a multiple of the chemical cell, and not least the extra degree of freedom with the magnetization as a vector instead of a scalar slows the iterative convergence, and allows for many more metastable states that must in general be sorted out.

In addition to treating commensurate magnetic cells, there exists a beautiful method [72, 155] to deal with non-commensurate spin density waves. It uses the fact that in case of so-called spin spirals, the magnetization is rotated in between different unit cells in the crystal, but is otherwise unchanged, at least in the absence of spin-orbit coupling. Since the magnitude of the magnetization is translationally invariant, one can introduce general symmetry operations that combine translations with spin rotations. This method has been used to calculate the observed helical or cycloidal spin density waves in, for instance, some transition metal systems [94] and in rare earth metals [133].

2.6 The LDA+U Method

The current approximations to the exchange-correlation functionals $E_{\rm xc}$ have clear limitations when it comes to systems with so-called correlated electrons, *e.g.* some transition metal oxides or rare earth compounds. In these d or f metal systems, the electronic states are close to localization and the Coulomb repulsion between the electrons within an open shell is of a completely different nature than in the homogeneous electron gas, upon which LSDA and GGA are based. This should, in principle, be remedied in a more exact version of the DFT. However, it is not known how to write the appropriate functionals in a standard orbital independent way. In the mean time, a completely different approach has been developed, which is to add a Hubbard like on-site repulsion on top of the usual Kohn-Sham Hamiltonian, [3, 4], *i.e.* to add,

$$E_U = U/2 \sum_{i \neq j} n_i n_j \,, \tag{2.33}$$

to the ordinary DFT xc energy, while subtracting a double counting term. Here n_i is the occupation number of orbital $i = \{m_\ell, \sigma\}$ in the relevant atomic shell ℓ. This method, known as LDA+U, was first developed to be able to cope with so-called Mott insulators, *i.e.* systems where LDA and GGA incorrectly predict a metallic state. By construction, the resulting potential is orbital dependent. This now allows for localization of occupied orbitals. However, since DFT also incorporates exchange and correlation in some sense, care has to be taken in order to correct for double counting. Unfortunately, there is not a unique way to make this correction. For instance, one can assume integer occupation numbers, which is relevant in the atomic limit, or equal non-integer occupation numbers for all orbitals, the so-called around mean field approach [145]. It has been observed that the different treatments of the double counting term can lead to qualitatively different physics, especially at intermediate values of U, and so it is important to note which scheme is being used. However, in general the effect of the Hubbard U in the LDA+U method is to drive the orbital occupations towards integers, and to favor insulating states over metallic ones. This may or may not be the correct physics in a given system. Especially in the case of metals, fluctuations not included in the LDA+U scheme can work against the tendency of the LDA+U method towards integer orbital occupations. As a result, the the LDA can provide a better description of the electronic structure than the LDA+U method, even in some moderately to strongly correlated metals. On the other hand, for correlated Mott-Hubbard insulators, the LDA description is unphysical, while the LDA+U approach provides a very reasonable description of the electronic structure.

The LDA+U method has evolved since its first suggestion. The most general versions use a parametrized screened Hartree-Fock interaction for electrons within one atomic shell [108, 185]. The renormalization of the bare exchange parameters is due to screening or correlations and depends strongly on the specific system. These parameters, *e.g.* U and J (in general there are $\ell + 1$ independent parameters for a shell of angular momentum ℓ), can be estimated by constrained DFT calculations, but often they are used as free parameters. In this Hartree-Fock like scheme, local density matrices, *i.e.* n_{ij}, are used instead of the occupation numbers in Eq. 2.33, which leads to a rotationally invariant formulation. This is what is generally implemented in codes.

The LDA+U method involves the identification of local atomic-like orbitals to which the non-LDA, orbital dependent, interaction is to be applied. Schick *et al.* [162] describe the implementation of the LDA+U approach in the context of the LAPW method. In their implementation, the Hubbard term is applied by projecting the bands onto the LAPW radial functions of selected angular momentum character (see Chapters 4 and 5, for an explanation of the radial functions), and using these projections to define the density matrix that then defines the U and J dependent parts of the Hamiltonian for the next self-consistent iteration.

Chapter 3

PLANEWAVE PSEUDOPOTENTIAL METHODS

A perusal of the electronic structure literature during the late 1970's and early 1980's when density functional calculations and particularly ab initio total energy methods were first showing their muscle, reveals that the field was largely dominated by planewave based pseudopotential methods. There is an interesting parallel today. The advent of ab initio molecular dynamics using the Car-Parrinello (CP) method [30] has resulted in a considerable leap in the capability of planewave based density functional methods, and application of these approaches has permitted the solution of numerous previously intractable problems. But even now, few substantial applications of these ideas to non-planewave based methods have been reported. Why?

We think that it is safe to say that the real reasons for the early dominance of planewave methods relate less to computational efficiency than to human efficiency. By this we mean that, because of the extreme simplicity of planewaves, the development and implementation of methods based on them is relatively straightforward. While other methods may have offered significant computational advantages, the extra human effort needed to develop them prevented their early emergence as methods of choice. We think that the situation is similar today, and see no fundamental reason why CP and related algorithms cannot be usefully exploited in non-planewave methods; just that it is more complicated and will take time. Now as then, practitioners of planewave methods are doing good physics while others are developing codes. However, when non-planewave total energy methods became available new areas of computational research were opened up in the d- and f-band materials, for example. Even today, planewave based CP methods have made little impact on important problems, such as the dynamics of high temperature superconductors and other correlated materials. Thus it seems quite reasonable to expect that as non-planewave CP-like codes are developed, new vistas will again be opened up, though to what extent remains to be seen.

As mentioned, this book has two purposes, one of which is to stimulate work aimed at combining the LAPW method with CP-like algorithms. In order to do this, we begin by reviewing some aspects of planewave pseudopotential and CP methods. This chapter is therefore intended to lay the necessary ground work for what follows; by no means is it intended as a comprehensive review of planewave based methods. Rather, readers seeking more detail are referred to one of the many excellent reviews already in the literature. We would particularly recommend the reviews of Payne *et al.* [139] (for CP and related methods), and also those of Pickett [148] and Cohen [35], as well as the papers in Ref. [40] for planewave pseudopotential methods in general.

3.1 Why Planewaves

Bloch's theorem, which starts with the periodicity of the crystal lattice, defines the crystal momentum **k** as a good (conserved) quantum number and also gives the boundary condition for the single particle wavefunctions, $\varphi_{\mathbf{k}}$. This is

$$\varphi_{\mathbf{k}}(\mathbf{r}+\mathbf{R}_L) = e^{i\mathbf{k}\cdot\mathbf{R}_L}\varphi_{\mathbf{k}}(\mathbf{r}), \tag{3.1}$$

were $\mathbf{R}_L$ is a direct lattice vector. The most general solution that satisfies this boundary condition is

$$\begin{aligned}\varphi_{\mathbf{k}}(\mathbf{r}) &= e^{i\mathbf{k}\cdot\mathbf{r}}\sum_{\mathbf{G}} c_{\mathbf{G}}(\mathbf{k})e^{i\mathbf{G}\cdot\mathbf{r}} \\ &= e^{i\mathbf{k}\cdot\mathbf{r}}w(\mathbf{k},\mathbf{r}),\end{aligned} \tag{3.2}$$

where the **G** are reciprocal lattice vectors. However, planewaves are diagonal in the momentum, **p**, and any powers of the momentum. This means that they are eigenfunctions of the kinetic energy operator $(\mathbf{p}^2/2m)$. Thus we have the solution for the empty lattice (constant potential, which we may choose to be zero). The band energies in Rydberg units are $(\mathbf{k}+\mathbf{G})^2$ and the wavefunctions are $e^{i(\mathbf{k}+\mathbf{G})\cdot\mathbf{r}}$ modulo a normalization. Now suppose that a small periodic perturbation, $\Delta V(\mathbf{r})$ is added to the Hamiltonian. Then the wavefunctions will no longer be pure planewaves, but mixtures. However, because of the energy denominators that occur in perturbation expansions, the mixture will involve mainly the nearby bands. Thus if our interest is in the lowest few bands, it will not be necessary to include a large number of additional planewaves provided that $\Delta V(\mathbf{r})$ is weak. Planewaves are ideally suited to the description of this nearly free electron (NFE) situation.

Solids, however, are made up of electrons and nuclei interacting strongly through the Coulomb potential. Nonetheless, according to Fermi liquid theory (FLT) the electronic excitations near the Fermi energy in metals behave as if they were independent particles. These quasiparticles have renormalized masses and intensities but are non-interacting at sufficiently low energies. Thus, besides its foundational role in the theory of metals, FLT provides guidance for band structure methods as well. That is, that we may expect the Hartree and exchange correlation potentials representing the interaction with the other valence electrons to be weaker than the bare Coulomb interactions would suggest, and perhaps weak enough to justify an NFE based approach in many cases.

This leaves the strong interactions with the core electrons and the nuclei to contend with. In most cases, the core electrons are quite strongly bound, and do not respond effectively to motions of the valence electrons. Thus they may be regarded as essentially fixed.

This is the essence of the pseudopotential approximation: The strong core potential (*n.b.* this does not mean $-Ze^2/r$ but rather includes as well the Hartree potential due to the core charge as well as a component of the exchange-correlation potential related to the valence-core interaction) is replaced by a pseudopotential, whose ground state wavefunction φ^{PS} mimics the all electron valence wavefunction outside a selected core radius. In this way, both the core states and the orthogonalization wiggles in the valence wavefunctions are removed. For many elements, the resulting pseudo-wavefunctions φ^{PS} are quite smooth and may be well represented using only low $|\mathbf{G}|$ planewaves. Thus planewaves become a simple and reasonably efficient basis for the pseudo-wavefunctions in these cases, and this underlies their popularity. Of course, there is a price to be paid. This is the need to generate and use a pseudopotential, rather than the actual crystal potential. In fact, much of the complexity of the method is transferred from the calculation itself to

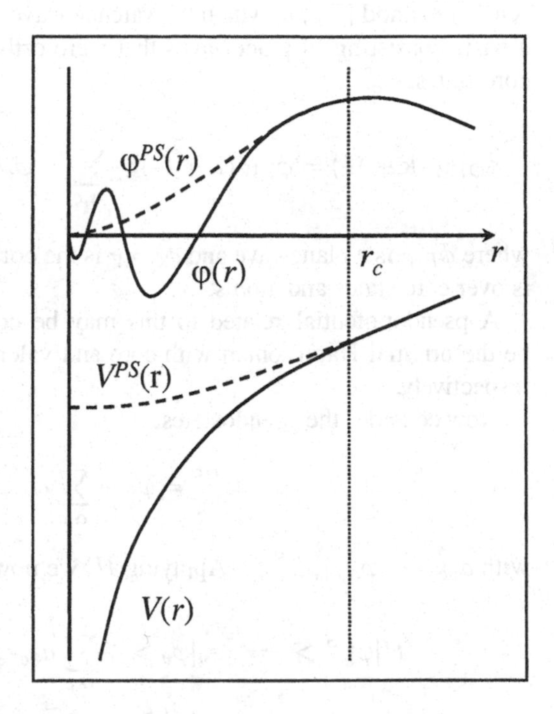

Figure 3.1. Schematic illustration of the replacement of the all-electron wavefunction and core potential by a pseudo-wavefunction and pseudopotential.

the generation of the pseudopotential. Further, the core states are fixed in an atomic reference configuration. The frozen core approximation, as it is called, is generally reliable, but breaks down for some elements with extended core states.

3.2 Pseudopotentials

The Phillips-Kleinman Construction

The pseudopotential approach originated with the orthogonalized planewave (OPW) method [72], in which the valence wavefunctions were expanded using a basis consisting of planewaves that were orthogonalized to the lower lying core states, φ_c.

$$\phi_{OPW}(\mathbf{k}+\mathbf{G}) = \phi_{PW}(\mathbf{k}+\mathbf{G}) - \sum_{\alpha,c} < \varphi_c|\phi_{PW}(\mathbf{k}+\mathbf{G}) > \varphi_{\alpha,c}, \quad (3.3)$$

where ϕ_{PW} is a planewave and ϕ_{OPW} is the corresponding OPW, and the sum is over core states and atoms.

A pseudopotential related to this may be constructed as follows: Let H be the original Hamiltonian with core and valence wavefunctions, φ_c and φ_v respectively.

Now consider the pseudostates,

$$\varphi_v^{PS} = \varphi_v + \sum_{\alpha,c} a_{vc}\varphi_{\alpha,c}, \quad (3.4)$$

with $a_{vc} =< \varphi_{\alpha,c}|\varphi_v^{PS} >$. Applying H, we now obtain

$$\begin{aligned} H|\varphi_v^{PS} > &= \varepsilon_v|\varphi_v > + \sum_{\alpha,c} a_{vc}\varepsilon_c|\varphi_{\alpha,c} > \\ &= \varepsilon_v|\varphi_v^{PS} > + \sum_{\alpha,c} a_{vc}(\varepsilon_{\alpha,c} - \varepsilon_v)|\varphi_{\alpha,c} > \end{aligned} \quad (3.5)$$

where $\varepsilon_{\alpha,c}$ and ε_v are the core and valence eigenvalues, respectively. Thus, using the definition of a_{vc},

$$[\, H + \sum_{\alpha,c}(\varepsilon_v - \varepsilon_{\alpha,c})|\varphi_{\alpha,c} >< \varphi_{\alpha,c}|\,]\varphi_v^{PS} = \varepsilon_v^{PS}\varphi_v^{PS}. \quad (3.6)$$

Therefore the pseudostates satisfy a Schrodinger-like equation with an additional contribution, V^R to the Hamiltonian.

$$V^R = \sum_{\alpha,c}(\varepsilon_v - \varepsilon_{\alpha,c})|\varphi_{\alpha,c} >< \varphi_{\alpha,c}|, \quad (3.7)$$

where V^R differs from a normal potential term in that it is energy dependent through ε_v. Adding V^R to the original potential, V, contained in the Hamiltonian, yields the Phillips-Kleinman [146] pseudopotential, V^{PK},

$$V^{PK} = V + V^R. \tag{3.8}$$

Outside the core region, V^{PK} becomes equal to V as the core wavefunctions vanish. Thus, there is some radius, r_c around an atom beyond which the contribution of that atom to V^R is negligible. Moreover, the construction is linear in the sense that there is a separate and independent additive contribution from each atom α. Most importantly, because of the additional repulsive contribution in the core, the pseudopotential is generally much weaker than the original potential, resulting in reasonable convergence of planewave expansions of the pseudo-wavefunctions.

Norm Conserving Pseudopotentials

The sophistication and efficacy of pseudopotentials have evolved considerably since the Phillips-Kleinman construction. This evolution has been driven, at least in large part, by the following goals: (1) First of all the pseudopotential should be as soft as possible, meaning that it should allow expansion of the valence pseudo-wavefunctions using as few planewaves as possible; (2) Secondly, it should be as transferable as possible (meaning that a pseudopotential generated for a given atomic configuration should reproduce others accurately), thereby helping to assure that the results will be reliable in solid state applications, where the crystal potential is necessarily different from an atomic potential; and (3) the pseudo-charge density (the charge density constructed using the pseudo-wavefunctions) should reproduce the valence charge density as accurately as possible. The concept of norm-conservation [192, 187] represented an important advance in reconciling these conflicting goals. With norm conserving pseudopotentials, the pseudo-wavefunctions (and potential) are constructed to be equal to the actual valence wavefunctions (and potential) outside some core radius, r_c. Inside r_c, the pseudo-wavefunctions differ from the true wavefunctions, but the norm is constrained to be the same. That is,

$$\int_0^{r_c} \mathrm{d}r\ r^2 \varphi^{PS\,*}(r)\varphi^{PS}\,(r) = \int_0^{r_c} \mathrm{d}r\ r^2 \varphi^*(r)\varphi(r), \tag{3.9}$$

where the wavefunctions refer to the atomic reference state and spherical symmetry is enforced. Of course, the wavefunction and eigenvalue are different for different angular momenta, ℓ and this implies that the pseudopotential should also be ℓ dependent. Pseudopotentials of this type are often called semi-local.

One measure of transferability is provided by the logarithmic derivatives at r_c of the all electron and pseudo-wavefunctions, φ and φ^{PS}, respectively. The imposed equality of these for $r \geq r_c$ ensures that the logarithmic derivatives at r_c are also equal for the atomic reference configuration.

$$\frac{1}{\varphi^{PS}(r_c, E)} \frac{\mathrm{d}\varphi^{PS}(r_c, E)}{\mathrm{d}r} = \frac{1}{\varphi(r_c, E)} \frac{\mathrm{d}\varphi(r_c, E)}{\mathrm{d}r}, \quad (3.10)$$

where E is the reference energy; Eqn. 3.10 holds exactly for E equal to the atomic reference eigenvalue. The transferability is then defined by the range of E over which Eqn. 3.10 holds adequately. However (using Green's theorem; see Shaw and Harrison [161], for details),

$$-\frac{\partial}{\partial E} \frac{\partial}{\partial r} \ln \varphi(r_c, E) = \frac{1}{r_c^2 \varphi^*(r_c, E) \varphi(r_c, E)} \int_0^{r_c} \mathrm{d}r \; r^2 \varphi^*(r, E) \varphi(r, E). \quad (3.11)$$

Therefore, imposition of norm-conservation ensures not only that the logarithmic derivative of the pseudo- and all-electron wavefunction match at the reference energy, but also that the first derivative with respect to E matches as well. Thus, the difference between the pseudo- and all-electron logarithmic derivative is second order in the deviation from the reference, and this helps ensure transferability for norm-conserving pseudopotentials. Methods for constructing soft-core, semi-local norm conserving pseudopotentials were developed by Hamann *et al.* [67] and further refined by Bachelet *et al.* [6] (BHS) who tabulated accurate pseudopotentials for all the elements in the periodic table. Kerker [80] has presented an alternate approach, which yields pseudopotentials of comparable quality. This approach uses simple analytic representations for the pseudo-wavefunctions inside r_c. This was

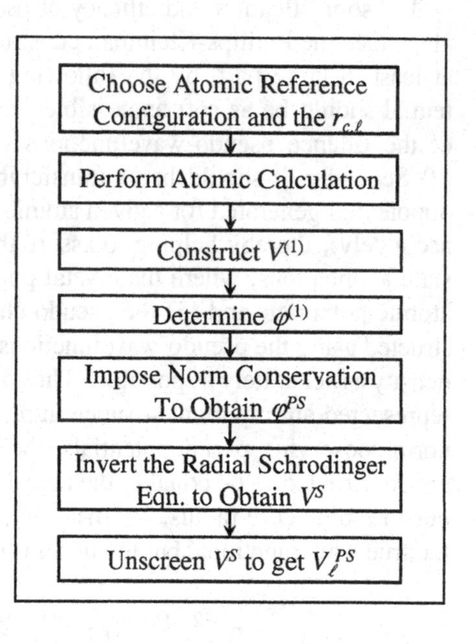

Figure 3.2. Bachelet, Hamann, Schluter procedure for generating norm-conserving pseudopotentials.

further refined by Troullier and Martins [193, 194], whose method is commonly used at present to generate norm-conserving pseudopotentials for practical calculations.

The tabulation of BHS was quite significant because it divorced ab initio calculations from pseudopotential generation and thereby substantially lowered the barrier to entry of the former field. This is no doubt partially responsible for the subsequent increase in activity in planewave pseudopotential studies of solids. The basic BHS procedure (Fig. 3.2) for constructing norm-conserving pseudopotentials is as follows (non-relativistic; see BHS for the relativistic case):

1 First an atomic reference configuration is selected and the atomic wavefunctions, eigenvalues, charge density and potential are calculated using an sphericalized atomic program (*e.g.* Liberman *et al.* [107]).

2 Core radii, $r_{c,\ell}$ are selected for each ℓ; $r_{c,\ell}$ must be between the outermost node in the all-electron valence wavefunction and its final extremum. Small values of $r_{c,\ell}$ result in hard (many planewaves needed to represent the pseudo-wavefunctions) but highly transferable pseudopotentials; large $r_{c,\ell}$ yield softer but less transferable pseudopotentials. Values of $r_{c,\ell}$ too close to the outermost node lead to numerical instabilities.

3 A first step pseudopotential, $\hat{V}_\ell(r)$ is constructed by cutting off the singularity in the all-electron atomic potential.

$$\hat{V}_\ell(r) = V(r)[1 - f(\frac{r}{r_{c,\ell}})] + c_\ell f(\frac{r}{r_{c,\ell}}), \tag{3.12}$$

where $f(x)$ is a monotonic function with $f(0) = 1, f(\infty) = 0$ and cutting off rapidly about $x = 1$. The constant c_ℓ is adjusted so that the lowest solution of the radial Schrodinger's equation with this potential has an eigenvalue equal to the atomic valence eigenvalue. Since $\hat{V}_\ell(r)$ and $V(r)$ are the same for r beyond the cutoff, the solution of the radial Schrodinger's equation with potential $\hat{V}_\ell(r)$ (*i.e.* $\varphi^{(1)}(r)$) is the same as that with $V(r)$ (that is, $\varphi(r)$), to within a multiplicative constant in this region.

4 Norm conservation is now imposed by adding a correction to $\varphi^{(1)}(r)$ in the core region.

$$\varphi_\ell^{PS}(r) = \gamma_\ell[\varphi^{(1)}(r) + \delta_\ell g_\ell(r)], \tag{3.13}$$

where γ_ℓ is the ratio $\varphi/\varphi^{(1)}$ outside $r_{c,\ell}$, δ is the parameter chosen to enforce norm conservation and $g_\ell(r)$ is a regular function ($\propto r^{\ell+1}$ for small r) that vanishes rapidly for $r > r_c$. BHS used $g_\ell(r) = r^{\ell+1} f(r/r_{c,\ell})$.

5 The screened pseudopotential V_ℓ^s (the potential that yields φ_ℓ^{PS} with the correct eigenvalue) is then obtained by numerical inversion of Schrodinger's equation.

6 Next, the pseudopotential is obtained by removing Hartree and exchange-correlation contributions, due to the total valence pseudocharge, from V_ℓ^s.

Finally, the accuracy, transferability and hardness of the pseudopotential, V_ℓ^{PS}, are tested by comparing all-electron and pseudopotential atomic calculations for several configurations.

Several modifications to the BHS and Kerker formulations have been developed that improve the resulting pseudopotentials, both in terms of transferability and in terms of hardness [196, 163, 152, 193, 93]. Basically, these modifications exploit the flexibility in the choice of the pseudo-wavefunction (and therefore the pseudopotential as well) within r_c. Further, Louie *et al.* [110] have developed an extended un-screening procedure (step 6) in which an exchange-correlation contribution due to the sum of the valence pseudocharge and a smooth stand-in for the core charge is removed from V^s. The core stand-in is included with the valence pseudocharge in calculating exchange-correlation contributions to the potential and energy when using the resulting pseudopotential. In this way, the non-linear core valence coupling present in the LDA may be included in pseudopotential calculations. This can greatly improve results for transition metals, especially when a very accurate treatment of the d states is needed, as in describing magnetism [34, 149].

Use of Norm Conserving Pseudopotentials with Planewaves

As mentioned, norm-conserving pseudopotentials generated by the BHS and related procedures have a semi-local form, since a different $V_\ell^{PS}(r)$ is generated for each atomic ℓ value. Thus,

$$V^{PS} = \sum_\ell V_\ell^{PS} \hat{P}_\ell. \tag{3.14}$$

Here, $\hat{P}_\ell$ is an angular momentum projection operator and the sum is over all ℓ. Fortunately, the $V_\ell^{PS}(r)$ converge rapidly as ℓ increases. Therefore, we may write

$$V^{PS} = V^{LOC}(r) + \sum_{\ell=0}^{\ell_{max}} \tilde{V}_\ell^{PS} \hat{P}_\ell, \tag{3.15}$$

where $V^{LOC}(r)$ is a local potential and ℓ_{max} is typically 1 or 2. Alternately, as is sometimes done, ℓ_{max} may be set one higher, and the $\ell = 0$ component taken as the local potential. In the following, the '$\tilde{}$' will be dropped, and the V_ℓ^{PS} will denote simply the non-local components of the pseudopotential.

With a planewave basis, the contributions to the matrix elements from the semi-local components of the pseudopotential (see Ihm *et al.* [75], for details) are:

$$\frac{1}{\Omega}\int \mathrm{d}^3\mathbf{r}\ \mathrm{e}^{-i(\mathbf{k}+\mathbf{G})\cdot\mathbf{r}}\ V_\ell^{PS}(r)\ \hat{P}_\ell\ \mathrm{e}^{-i(\mathbf{k}+\mathbf{G}')\cdot\mathbf{r}} = \\ \frac{2\ell+1}{\Omega}4\pi P_\ell(\cos\gamma)\int \mathrm{d}r\ r^2 V_\ell^{PS}(r) j_\ell(|\mathbf{k}+\mathbf{G}|r) j_\ell(|\mathbf{k}+\mathbf{G}'|r), \tag{3.16}$$

where

$$\cos\gamma = \frac{(\mathbf{k}+\mathbf{G})\cdot(\mathbf{k}+\mathbf{G}')}{|\mathbf{k}+\mathbf{G}|\,|\mathbf{k}+\mathbf{G}'|}, \tag{3.17}$$

the P_ℓ are Legendre polynomials, and for simplicity the origin has been taken at the atom (a structure factor is introduced if the atom is displaced from the origin). Although computable, this form has a some unpleasant aspects from a numerical point of view. First of all, because of the non-locality, the matrix elements depend not just on the difference, $(\mathbf{k}+\mathbf{G})-(\mathbf{k}+\mathbf{G}') = \mathbf{G}-\mathbf{G}'$, but separately on $(\mathbf{k}+\mathbf{G})$ and $(\mathbf{k}+\mathbf{G}')$. This means that if there are n planewaves in the basis, in general $n(n+1)/2$ independent terms need to be calculated instead of approximately $8n$ (*n.b.* the difference will have twice the maximum wavevector as the basis), as would have been the case for a local potential. This is quite significant for calculations involving large unit cells with many atoms; the number of planewaves is proportional to the volume of the cell, and thus the number of atoms, and each atom contributes separately to the matrix elements. Thus the computational work in setting up the Hamiltonian matrix using this formalism scales as the third power of the number of atoms as compared to the second power for a local potential. Secondly, the form itself is not efficacious for computations because of the complex dependence on the wavevectors inside the integral. This integral has to be evaluated (in practice two dimensional interpolations may be used) for each of the $n(n+1)/2$ pairs of wavevectors. Finally, the matrix elements depend not only on $\mathbf{G}$ and $\mathbf{G}'$, but also on $\mathbf{k}$ and therefore must be computed separately for each $\mathbf{k}$ point in the irreducible part of the Brillouin zone. This is in contrast to a local potential for which the matrix elements are $\mathbf{k}$ independent. However, in spite of these complications, semi-local pseudopotentials of this type have been used extensively and with considerable success in calculations.

The Kleinman-Bylander Transformation

Kleinman and Bylander (KB) [83] noted the above difficulties in performing planewave calculations using semi-local pseudopotentials and developed a transformation that largely circumvents them. The transformation (non-relativistic; see KB for a relativistic treatment) begins by adding and subtracting a local function, V^K

$$\sum_{\ell} V_{\ell}^{PS}(r)\,\hat{P}_{\ell} = V^K(r) + \sum_{\ell m} |Y_{\ell m} > \delta V_{\ell}(r) < Y_{\ell m}|, \tag{3.18}$$

where the projection operator has been expressed using spherical harmonics, and $\delta V = V^K + V_{\ell}^{PS}$. Next a non-local pseudopotential, V^{NL} is used to replace the last (semi-local) term, V^{SL}.

$$V^{NL} = \sum_{\ell m} \frac{|\delta V_{\ell}\, \varphi_{\ell m}^{PS} >< \varphi_{\ell m}^{PS} \delta V_{\ell}|}{< \varphi_{\ell m}^{PS} | \delta V_{\ell} | \varphi_{\ell m}^{PS} >}, \tag{3.19}$$

where φ_{ℓ}^{PS} is the pseudo-wavefunction, including the angular dependence, for the reference state. With this choice it is readily shown that

$$V_{\ell}^{NL}\, |\varphi_{\ell}^{PS} >= V_{\ell}^{SL}\, |\varphi_{\ell}^{PS} >, \tag{3.20}$$

again for the reference state and with the V_{ℓ}^{SL} being the ℓ components of V^{SL}. Thus, for the reference state at least, these two pseudopotentials are equivalent. KB and others have performed extensive numerical tests showing that, using this procedure with a careful choice of V^K, very high quality pseudopotentials may be obtained. Further, with this form the matrix elements become much simpler. This is because

$$< \mathbf{G}|V|\mathbf{G}' >= \sum_j < \mathbf{G}|v_j >< v_j|\mathbf{G}' >, \tag{3.21}$$

if

$$V = \sum_j |v_j >< v_j|, \tag{3.22}$$

which is the form of the KB pseudopotential (Eqn. 3.19). This separation represents a substantial improvement over the semi-local form from a computational point of view. This is because, with this form, the matrix elements $< \mathbf{G}|v_j >$ may be precalculated and the products needed for the matrix elements evaluated quickly. In fact, the number of integral evaluations with the KB transformation scales as the second power of the number of atoms rather than the third as in the semi-local approach. Of course, the total work in setting up the Hamiltonian matrix still scales as the third power of the number of atoms in the unit cell (n^2 matrix elements, each containing a sum over atoms). An alternate approach is to operate the projectors in real space form, usually by simply calculating the values on a the real space grid used to represent the wave-functions. While this improves the scaling, it introduces numerical problems unless great care is taken to accurately represent the projectors [136].

Ultrasoft Pseudopotentials (Formalism)

Vanderbilt and co-workers [197, 98, 99] proposed a radical departure from the concept of norm-conservation, as it is discussed above. In their approach, the pseudo-wavefunctions are required to be equal to the all-electron wave-functions outside r_c, as with norm-conserving pseudopotentials, but inside r_c they are allowed to be as soft as possible; the norm-conservation constraint is removed to accomplish this. Although this introduces some complications, it can greatly reduce the planewave cutoff needed in calculations, particularly since quite large values of r_c can be used in their scheme. The complications that result are two-fold. First of all, since the pseudo-wavefunctions are equal to the all-electron wavefunctions (and therefore have the same norm) in the interstitial, but do not have the same norm inside r_c they are necessarily not normalized. This introduces a non-trivial overlap into the secular equation. In fact, the overlap turns out (see below) to be non-diagonal. Secondly, the pseudo-charge density is not obtained by computing $\sum \varphi^* \varphi$ as with norm conserving pseudopotentials; among other things this would yield the wrong total charge. Rather, an augmentation term needs to be added in the core region. A third, but less important, complication is that by relaxing the norm conservation, the resulting pseudopotentials can become less transferable. However, Vanderbilt pseudopotentials were proposed for use in large scale calculations, for which the cost of generating pseudopotentials is negligible compared with the cost of the calculations. Accordingly, it is quite feasible to recalculate the pseudopotential as the configuration evolves during the course of the calculation [197].

In Vanderbilt's ultrasoft pseudopotential approach the total energy is written as,

$$E = \sum_{occ} < \varphi_j | T + V^{NL} | \varphi_j > + \int d^3\mathbf{r}\, V^L(\mathbf{r}) \rho(\mathbf{r}) + \frac{1}{2} \int d^3\mathbf{r} d^3\mathbf{r}' \frac{\rho(\mathbf{r})\rho(\mathbf{r}')}{|\mathbf{r}-\mathbf{r}'|} + E_{xc}[\rho] + E_{ii}, \tag{3.23}$$

where T is the kinetic energy operator, V^L is the local component of the pseudopotential, V^{NL} is the non-local Vanderbilt pseudopotential, the φ_j are the pseudo-wavefunctions and the other terms are as before (see section 2.2). A fully non-local separable from is used for V^{NL},

$$V^{NL} = \sum_{mn} D^{(0)}_{nm} |\beta_n >< \beta_m|, \tag{3.24}$$

where, for simplicity, only one atom is considered, and the pseudopotential is characterized by the functions, β_m, the coefficients $D^{(0)}_{nm}$ and the local component $V^L(r)$. The β_m are represented in an angular expansion, *i.e.* spherical harmonics multiplied by radial functions (typically one or two are used for each ℓm combination). The radial functions vanish outside r_c.

As mentioned, the pseudo-charge density ρ is given by the square of the pseudo-wavefunctions plus an augmentation inside the spheres.

$$\rho(\mathbf{r}) = \sum_{occ} [\varphi^*_j(\mathbf{r})\varphi_j(\mathbf{r}) + \sum_{mn} Q_{nm}(\mathbf{r}) < \varphi_j|\beta_n >< \beta_m|\varphi_j >], \tag{3.25}$$

where the $Q_{nm}(\mathbf{r})$ are local functions determined during the generation of the pseudopotential.

Applying the variational principle to Eqns. 3.23 to 3.25, the secular equation is

$$H|\varphi_j >= \varepsilon_j S|\varphi_j >, \tag{3.26}$$

with

$$H = T + V_{xc}(\mathbf{r}) + V_H(\mathbf{r}) + V^L(\mathbf{r}) + \sum_{mn} D_{nm}|\beta_n >< \beta_m|, \tag{3.27}$$

and

$$S = \mathbf{1} + \sum_{mn} q_{nm}|\beta_n >< \beta_m|, \tag{3.28}$$

where $\mathbf{1}$ denotes the identity operator and

$$q_{nm} = \int_\alpha \mathrm{d}^3\mathbf{r}\, Q_{nm}(\mathbf{r}), \tag{3.29}$$

with the integral over the sphere defined by r_c. The D_{nm} are the $D^{(0)}_{nm}$ with a screening term.

$$D_{nm} = D^{(0)}_{nm} + \int_\alpha \mathrm{d}^3\mathbf{r}\, V(\mathbf{r}) Q_{nm}(\mathbf{r}), \tag{3.30}$$

where V denotes the local potential, given by the local pseudopotential plus the exchange correlation and Hartree potentials.

As will be discussed in Chapter 6, the LAPW method may be transformed to appear very similar to a planewave method with a pseudopotential that has strong similarities to the Vanderbilt pseudopotential.

Ultrasoft Pseudopotentials (Generation)

The Vanderbilt scheme [197, 99] for generating ultrasoft pseudopotentials begins with all-electron atomic calculations in some reference configuration. For each angular momentum, a set (typically 1-3) of reference energies, $E_{\ell j}$, is chosen spanning the range over which band states will be calculated. The radial Schrodinger equation is then solved within r_c at each $E_{\ell j}$, yielding regular solutions $\varphi_{\ell m j}(\mathbf{r}) = u_{\ell j}(r) Y_{\ell m}(\hat{\mathbf{r}})$. For each $\{\ell m j\}$, a smooth pseudowavefunction, $\phi_{\ell m j}(\mathbf{r}) = \hat{u}_{\ell j}(r) Y_{\ell m}(\hat{\mathbf{r}})$ is generated subject only to the constraint that it match smoothly onto $\varphi_{\ell m j}$ at r_c. Similarly, a smooth local potential, V^L, that matches the all-electron potential outside r_c is determined. Next the orbitals,

$$|\chi_{\ell m j}> = \left[E_{\ell j} - T - V^L(\mathbf{r})\right] |\phi_{\ell m j}>, \tag{3.31}$$

are constructed. Since ϕ and V^L are equal to φ and the all-electron potential, respectively, outside r_c, and φ satisfies Schrodinger's equation at $E_{\ell j}$, χ is zero outside r_c. Now the $Q_{nm}(\mathbf{r})$ can be constructed, since we know that it must account for the difference between the true charge density and $\phi^*\phi$.

$$Q_{nm}(\mathbf{r}) = \varphi^*_n(\mathbf{r})\varphi_m(\mathbf{r}) - \phi^*_n(\mathbf{r})\phi_m(\mathbf{r}), \tag{3.32}$$

where n and m run over the $\{\ell m j\}$. In practice, a smoothing may be applied to Q_{nm} in order to facilitate use with planewave representations of the charge density. If this is done, the smoothing is constructed to preserve the moments of the original Q_{nm}. We may similarly construct the $|\beta_n>$.

$$|\beta_n> = \sum_m (\mathbf{B}^{-1})_{mn} |\chi_m>, \tag{3.33}$$

with $\mathbf{B}_{nm} = <\phi_n|\chi_m>$.

The remaining components of the pseudopotential, V^L and D_{nm}, are determined by the identity,

$$[\,T+V+\sum_{nm} D_{nm}|\beta_n><\beta_m|\,]\;|\phi_n>=$$
$$E_n\,[\mathbf{1}+\sum_{nm} q_{nm}|\beta_n><\beta_m|\,]\;|\phi_n>, \tag{3.34}$$

which holds if

$$D_{nm}=\mathbf{B}_{nm}+E_m q_{nm}. \tag{3.35}$$

Finally, the D_{nm} are unscreened using Eqn. 3.30 to determine the $D^{(0)}_{nm}$, and the Hartree and exchange correlation contributions are subtracted from V to obtain V^L.

An interesting feature of this pseudopotential is the fact that as the self-consistent iterations proceed, the contribution of the augmenting charge inside the sphere changes along with the wavefunctions. This charge contributes to the potential used in the Kohn-Sham equations. Since this contribution may be described as being part of the pseudopotential, one may regard the pseudopotential as evolving during the calculation. In any case, the evolution of the augmenting charge and its contribution to the potential during the calculation allows relatively large values of r_c to be used in the Vanderbilt construction. This yields very soft pseudopotentials, without sacrificing the accuracy of the calculation.

3.3 Introduction to the Car-Parrinello Method

Overview

In standard density functional approaches, the degrees of freedom associated with the electronic wavefunctions (ordinarily, the coefficients of the basis functions, which comprise the eigenvectors) are viewed quite differently from those associated with the nuclear coordinates. In particular, a hierarchical view is taken in the sense that the former are treated as functions of the latter. Further, the computational methods used to determine these two sets of coordinates are quite distinct. The electronic degrees of freedom are determined by self-consistent solution of the Kohn-Sham equations for a fixed set of nuclear positions, and the latter by varying the nuclear coordinates using information from total energy and usually force calculations. In contrast, the CP method treats the electronic and nuclear degrees of freedom on the same footing using methods related to classical molecular dynamics. This is done by simulating a fictitious classical dynamical system containing both sets of degrees of freedom.

Car and Parrinello [30] begin with the following fictitious Lagrangian for the electronic eigenvector coefficients.

$$L = \sum_j \frac{1}{2}\mu < \dot{\varphi}_j|\dot{\varphi}_j > -E[\{\varphi\},\{\mathbf{r}\}], \qquad (3.36)$$

where the sum is over occupied states, E is the total electronic energy, μ is a fictitious mass, the $\mathbf{r}$ are the nuclear coordinates, the φ are the electronic wavefunctions and the $\dot{\varphi}$ are their time derivatives. This Lagrangian is in a form that allows the simulation of a fictitious electronic dynamics. The addition of a term containing the kinetic energy associated with the nuclear degrees of freedom yields a Lagrangian in a form amenable to classical simulation of both nuclear and fictitious electronic motions.

Unfortunately, this is not the whole story, since the orbitals entering Kohn-Sham electronic structure theory are not arbitrary but must rather be orthonormal. This means that the variations, $\dot{\varphi}\delta t$, are constrained. In the CP formalism, these constraints are incorporated using Lagrange multipliers Λ_{ij}. This yields a modified Lagrangian,

$$L = \sum_j \frac{1}{2}\mu < \dot{\varphi}_j|\dot{\varphi}_j > -E[\{\varphi\},\{\mathbf{r}\}] + \sum_{ij}\Lambda_{ij}[\int_\Omega \mathrm{d}^3\mathbf{r}\, \varphi_i^*\varphi_j - \delta_{ij}], \quad (3.37)$$

where the integral is over the unit cell.

Inserting this expression for L into the Lagrange equations of motion yields the dynamical equations,

$$\mu\ddot{\varphi}_j = -H\varphi_j + \sum_i \Lambda_{ji}\varphi_i, \qquad (3.38)$$

where H is the usual Kohn-Sham single particle Hamiltonian. Car and Parrinello emphasize that the dynamics described by this Lagrangian are fictitious; density functional theory is valid only on the Born-Oppenheimer surface, *i.e.* only when the electronic degrees of freedom are stationary. Nonetheless, these equations do provide a useful way of solving the electronic structure problem.

Consider first the case where the nuclear coordinates are held fixed. In this case, the dynamics describe the trajectory of the electronic eigenvectors, starting with an initial guess, and with the Kohn-Sham Hamiltonian acting as a force. The electronic structure problem may then be solved by simulated annealing [82]. In this approach, a damping term is added to the equations of motion, so that, at each time step in the simulation, fictitious kinetic energy associated with the electronic degrees of freedom is removed. Since the Lagrangian depends on the energy functional for the current eigenvectors, the Kohn-Sham Hamiltonian appearing in the dynamical equations is updated at each time step. This amounts to recomputing the charge density and potential each time that the eigenvectors are moved. When all available energy is removed in this way, the electronic structure problem is solved. In the limit in which the maximum amount of kinetic energy is removed at each time step, the simulated annealing procedure can be reduced to the method of steepest descents for minimizing a function, specifically the total energy (see Williams and Soler [206]). At each step, each eigenvector coefficient is moved in the direction of the force on it, by a distance depending on the force and time step, δt, and inversely on the fictitious electronic mass, μ. Because of these dependencies, judicious choices of δt and μ are needed. If δt is too short (or μ too large), the moves will be small, and many steps will be needed to reach the solution. On the other hand, if δt is too long (or μ too small), the integration of the dynamical equations will become unstable, and the solution may not be found at all.

A more general situation occurs when the nuclear coordinates are also to be determined. In this case, the nuclear degrees of freedom are simply included in the simulated annealing run. As such, the electronic eigenvector coefficients evolve under the action of the Kohn-Sham Hamiltonian, and the nuclear coordinates evolve with the Hellmann-Feynman forces due to the current electronic structure. When all available energy has been removed, both the electronic and nuclear degrees of freedom will be stationary, so that the nuclear coordinates will rest at a minimum (perhaps local) in the energy surface and the eigenvector coefficients will solve the electronic structure problem for that geometry. The trajectory that will be followed is determined by the simulated annealing algorithm, the initial values, and by δt, μ and the nuclear masses, M_i. For small ratios of μ/M_i, the electronic degrees of freedom will evolve more rapidly than the nuclear degrees of freedom; in the limit where the ratio goes to zero, the trajectory will follow the Born-Oppenheimer surface. As μ is increased, the calculation becomes more efficient in that the solution is found in fewer time steps, but the deviation from the Born-Oppenheimer surface also increases, and this can make the procedure unstable.

Finally, we note that the CP approach can be used not only to find zero temperature solutions of the Kohn-Sham equations, but also to perform *ab initio* molecular dynamics (see, for example, Galli *et al.* [47]). In this case, μ is chosen to yield a solution sufficiently close to the Born-Oppenheimer surface so that the dynamics of the nuclei approximate adequately the physical dynamics. In this way, finite temperature simulations can be done in which the coordinates evolve according to forces determined by the full density functional electronic structure.

Application to Planewave Electronic Structure Calculations

In the CP approach to electronic structure calculations, the eigenvectors are iteratively refined towards the solution using the dynamical equations. This refinement requires repeated evaluation of the operation of the Kohn-Sham Hamiltonian on the current eigenvectors. In general, both the number of eigenvectors (the number of bands, n_e) and the length of the eigenvectors (the number of basis functions, n_b) are extensive quantities, *i.e.* they are proportional to the number of atoms in the unit cell. If the operation proceeds by multiplying a precomputed Hamiltonian matrix (dimension $n_b \times n_b$) by each eigenvector, then the computational work in this step is at least $n_e n_b^2$, *i.e.* proportional to the number of atoms, N, to the third power. However, with a planewave basis set this operation can be performed much more efficiently. In particular, with this basis, the kinetic energy operator is diagonal in momentum space, while a local potential is diagonal in real space. The eigenvectors can be transformed efficiently between real and momentum space using fast Fourier transform (FFT) methods. Exploiting this to operate the potential in real space and the kinetic energy in momentum space, the computational cost of operating the Hamiltonian on an eigenvector is dominated by the $O(n_b \log_2(n_b))$ cost per band of the FFTs and as a result this part of the calculation scales with the number of atoms as $O(N^2 \log N)$. Although for large systems the total cost of a CP calculation scales as $O(N^3)$, because of the needed orthogonalizations of the eigenvectors (see *e.g.* Ref. [27], for details), the prefactor of the remaining $O(N^3)$ term is much smaller than in conventional diagonalization based methods. This is the reason why the CP method implemented using a planewave basis is so efficient, and also explains, at least in large part, why comparable gains are difficult to realize by the application of CP-like algorithms to non-planewave basis sets.

In the original CP procedure a global electronic mass, μ, is used in the simulation. As mentioned, a careful choice of this mass is important in obtaining the electronic structure in a reasonable number of time steps. Payne *et al.* [138] recognized that it is neither necessary nor desirable to treat all the electronic degrees of freedom on the same footing. In fact, they showed that the integration of the equations of motion in a planewave basis yields harmonic oscillator solutions; the oscillator frequency, ω, depends on the wavevector, $\mathbf{k} + \mathbf{G}$, and for large wavevectors is proportional to $|\mathbf{k} + \mathbf{G}|$. Taking account of this fact, they analytically integrate the equation of motion up to δt and thereby are able to use much larger values of the time step δt. This may be regarded as a kind of preconditioning.

As mentioned, an alternate view in the context of solving an electronic structure problem is that the essence of the CP method is to iteratively refine the eigenvectors (and corresponding charge density) in order to find the self-consistent electronic structure. From this point of view, the original formulation of the CP method uses a rather crude iterative procedure. The eigenvectors are just moved by some amount in a direction determined operating the Hamiltonian. However, a more efficient iterative approach might be preferred, because fewer iterations and therefore fewer evaluations of $H\varphi$ would be required. In fact, alternative efficient iterative approaches are known (see Wood and Zunger [210], and Payne *et al.* [139], for details). In general, these schemes contain orthogonalization steps, scaling as $O(N^3)$, as in the original CP method, and therefore have the same computational scaling with system size. Their advantage is that convergence is obtained in fewer steps. A common feature of iterative diagonalization schemes applicable to the electronic structure problem is that they use moves that are derived from computation of the residual vectors, $\mathbf{R}$.

$$\mathbf{R}_j = (\mathbf{H} - \lambda_j \mathbf{S})\varphi_j, \tag{3.39}$$

where $\mathbf{S}$ is the overlap matrix, which is diagonal in a planewave basis set, λ_j is the expectation value, and bold face is used to emphasize that the matrix - vector operations involved.

$$\lambda_j = \frac{< \varphi_j | H | \varphi_j >}{< \varphi_j | \varphi_j >}. \tag{3.40}$$

The residuals have the property that they are zero for the exact eigenvectors. Accordingly, an alternate view of the diagonalization process is minimization of the residuals.

Eqn. 3.39 provides a prescription (inverse iteration) for computing eigenvectors. In particular, if λ were the true eigenvalue, then

$$\varphi_j' = \varphi_j - (\mathbf{H} - \lambda \mathbf{S})^{-1} \mathbf{R}_j, \tag{3.41}$$

would be the correct eigenvector (*n.b*, if λ is the correct eigenvalue then $\mathbf{H} - \lambda\mathbf{S}$ is singular and the inverse is to be understood as the inverse as obtained via singular value decomposition with φ' removed). Unfortunately, direct application of this by iterating Eqns. 3.40 and 3.41 is not practical in CP calculations since it would require inversion of the Hamiltonian matrix, which is an $O(n_b^3)$ process. It does, however, suggest one way of improving the original procedure. In particular, if a good guess, $\mathbf{D}$, for $(\mathbf{H} - \lambda\mathbf{S})^{-1}$ can be made, updating the eigenvectors with $-\mathbf{DR}$ should yield better convergence than updating with $-\mathbf{R}$. In planewave pseudopotential calculations, $\mathbf{H}$ is strongly diagonally dominant for large wavevectors because of the kinetic energy contribution. This suggests the following approximate form for $\mathbf{D}$,

$$\mathbf{D}_{\mathbf{G},\mathbf{G}'} = \delta_{\mathbf{G},\mathbf{G}'} f(|\mathbf{k} + \mathbf{G}|^2), \tag{3.42}$$

where $f(x) \propto 1/x$ for large x. The resulting updating procedure (preconditioned steepest descent) is considerably more efficient in planewave calculations than the original steepest descent approach. In modern planewave codes, direct preconditioned steepest descents minimization is often avoided in favor of more sophisticated minimization schemes particularly preconditioned conjugate gradients (see Press *et al.* [150] for a general discussion, and Payne *et al.* [139] for application to electronic structure calculations). However, as with the steepest descent and simulated annealing approaches, these methods benefit from preconditioning of the residuals. Unfortunately, little progress has been made in developing preconditioning schemes for situations in which the Hamiltonian is not strongly diagonally dominant. This is a serious issue in the application of these methods to non-planewave basis sets, such as the LAPW method and the APW plus local orbital method.

A related set of iterative diagonalization methods are derived from the discrete inversion in iterative subspace (DIIS) approach [210]. In these methods a Hamiltonian matrix is constructed, either in the space of the updated eigenvectors (using preconditioned steepest descents) or in the space of the current eigenvectors augmented by the preconditioned residuals, and this is diagonalized to obtain the new eigenvectors. The dimension of this second variational Hamiltonian is $n_e \times n_e$ in the former case and $2n_e \times 2n_e$ in the latter. Advantages of the DIIS scheme are the automatic orthogonalization of the eigenvectors and its stability against solutions in which one or more eigenvectors are missed. However, it should be emphasized that the issue of finding effective preconditioning is not avoided by using DIIS.

To summarize, the important factors that underlie the efficiency and power of the CP method are (1) the use of iterative diagonalization schemes, which allow simultaneous determination of the electronic eigenvectors, self-consistent charge density, and nuclear coordinates, and (2) the use of a planewave basis set for which $H\varphi$ and the Hellmann-Feymann forces can be evaluated efficiently. In order for similar gains to be obtained from the application of CP-like algorithms to non-planewave methods these conditions must be satisfied, *i.e.* it must be possible to evaluate $H\varphi$ and the forces efficiently and in a numerically stable fashion, and effective, well conditioned, iterative schemes for the eigenvectors must be available.

Chapter 4

INTRODUCTION TO THE LAPW METHOD

4.1 The Augmented Planewave Method

The LAPW method is fundamentally a modification of the original augmented planewave (APW) method of Slater [176, 178]. Thus, before embarking on an exposition of the LAPW method, we review the relevant aspects of the APW method and the motivation for its modification to the LAPW method. Further details about the APW method may be found in the book by Loucks [111], which also reprints several early papers that develop and use this method.

Slater, in his 1937 paper developing the APW method, clearly and concisely states the essence of the method and its motivation: Near an atomic nucleus the potential and wavefunctions are similar to those in an atom - they are strongly varying but nearly spherical. Conversely, in the interstitial space between the atoms both the potential and wavefunctions are smoother. Accordingly, space is divided into regions and different basis expansions are used in these regions: radial solutions of Schrodinger's equation inside non-overlapping atom centered spheres and planewaves in the remaining interstitial region (Fig. 4.1).

$$\varphi(\mathbf{r}) = \begin{cases} \Omega^{-1/2} \sum_{\mathbf{G}} c_{\mathbf{G}} e^{i(\mathbf{G}+\mathbf{k})\cdot\mathbf{r}} & \mathbf{r} \in I \\ \sum_{\ell m} A_{\ell m} u_\ell(r) Y_{\ell m}(\hat{\mathbf{r}}) & \mathbf{r} \in S, \end{cases} \tag{4.1}$$

where φ is a wavefunction, Ω is the cell volume, u_ℓ is the regular solution of

$$\left[-\frac{d^2}{dr^2} + \frac{\ell(\ell+1)}{r^2} + V(r) - E_\ell \right] r u_\ell(r) = 0. \tag{4.2}$$

Here, $c_{\mathbf{G}}$ and $A_{\ell m}$ are expansion coefficients, E_ℓ is a parameter, V is the spherical component of the potential in the sphere and Rydberg units have been assumed. The radial functions defined by Eqn. 4.2 are automatically orthogonal to any eigenstate of the same Hamiltonian that vanishes on the sphere boundary [2]. This is shown by noting that from Schrodinger's equation,

$$(E_2 - E_1) r u_1(r) u_2(r) = u_2 \frac{\mathrm{d}^2 \, r u_1(r)}{\mathrm{d}r^2} - u_1 \frac{\mathrm{d}^2 \, r u_2(r)}{\mathrm{d}r^2}, \tag{4.3}$$

where $u_1(r)$ and $u_2(r)$ are radial solutions at different energies E_1 and E_2. The overlap is constructed using this relation and integrating by parts; the surface terms vanish if either $u_1(r)$ or $u_2(r)$ vanish on the sphere boundary, while the other terms cancel.

Slater motivates these particular choices of functions by noting that planewaves are the solutions of Schrodinger's equation in a constant potential, while the radial functions are solutions in a spherical potential, provided that E_ℓ is equal to the eigenvalue. This approximation to the potential is crude but can be reasonable in some cases. In fact, the muffin-tin (MT) approximation, as it is called, was widely used in APW codes to elucidate properties of transition metals and compounds. The MT approximation is very good for close packed (fcc and ideal c/a *hcp*) materials. It is less good but still reasonable for *bcc* and related (e.g. CsCl structure) materials [43], and becomes increasingly less reliable as the site symmetry and coordination decrease.

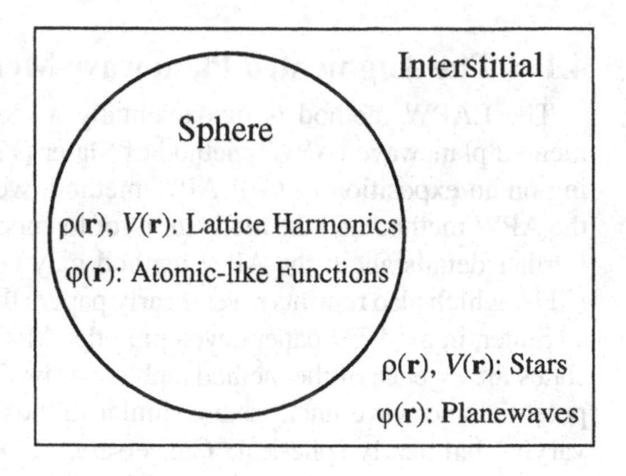

Figure 4.1. The dual representation of the APW and LAPW methods. Stars and lattice harmonics are symmetrized planewaves and spherical harmonics used to represent the density and potential.

There is one remaining point. That is that the dual representation defined by Eqn. 4.1 is not guaranteed to be continuous on the sphere boundaries, as it must be for the kinetic energy to be well defined. Accordingly, it is necessary to impose this constraint. In the APW method, this is done by defining the coefficients $A_{\ell m}$ in terms of the $c_{\mathbf{G}}$ through the spherical harmonic expansion of the planewaves. The coefficient of each ℓm component is matched at the sphere boundary. Thus after some algebra (see Chapter 5),

$$A_{\ell m} = \frac{4\pi i^{\ell}}{\Omega^{1/2} u_{\ell}(R)} \sum_{\mathbf{G}} c_{\mathbf{G}} j_{\ell}(|\mathbf{k}+\mathbf{G}|R) Y^{*}_{\ell m}(\mathbf{k}+\mathbf{G}), \tag{4.4}$$

where the origin is taken at the center of the sphere and R is the sphere radius. Thus, the $A_{\ell m}$ are completely determined by the planewave coefficients, $c_{\mathbf{G}}$ and the energy parameters, E_{ℓ}. These are the variational coefficients in the APW method. The individual functions, which are labeled by $\mathbf{G}$ and consist of single planewaves in the interstitial matched to radial functions in the spheres, are the augmented planewaves, or APWs.

If E_{ℓ} were taken as a fixed parameter, rather than a variational coefficient, the APW method would simply amount to the use of the APWs as a basis. This would result in a standard secular equation (Eqn. 2.17; the APWs are not orthogonal so there would be a non-trivial overlap matrix, $\mathbf{S}$). The solution of this secular equation would then yield the band energies and wavefunctions. Unfortunately, this is not a workable scheme. The APWs are solutions of the Schrodinger's equation inside the spheres, but only at the energy E_{ℓ}; they lack variational freedom to allow for changes in the wavefunction as the band energy deviates from this reference. Accordingly, E_{ℓ} must be set equal to the band energy. This means that the energy bands (at a fixed $\mathbf{k}$-point) cannot be obtained from a single diagonalization. Rather, it is necessary to solve the secular determinant as a function of energy and determine its roots – a much more computationally demanding procedure particularly for general $\mathbf{k}$-points, where the dimension of the secular equation cannot be folded down using symmetry, and systems with many bands (*n.b.* the number of bands grows with the number of atoms in the unit cell).

A further difficulty with the APW method is that it is hard (but not impossible [38, 43]), to extend it to use a general crystal potential, beyond the level of the warped muffin-tin approximation (general potential in the interstitial, but a spherical potential inside the spheres). This is because the optimum variational choice of E_{ℓ} is no longer given by setting E_{ℓ} to the band energy in this case. In particular, different bands will in general have different orbital characters inside the sphere (*e.g.* d_{z^2} *v.s.* $d_{x^2-y^2}$). However, in a non-spherical potential these orbitals experience different effective potentials, and these differ from the spherical average that is used to determine the radial function.

Another, less serious, difficulty with the APW method is the so called asymptote problem. In the expression (Eqn. 4.4) for the matching coefficient, $A_{\ell m}$, $u_{\ell}(R)$ appears in the denominator. However, there are in general values of the energy parameter, E_{ℓ} for which u_{ℓ} vanishes on the sphere boundary. At these energies the planewaves and radial functions become decoupled. In the vicinity of this asymptote the relation between the $A_{\ell m}$ and the $c_{\mathbf{G}}$ and thus the secular determinant are strongly varying. This leads to numerical difficulties when bands occur near an asymptote. In addition, care is needed in the vicinity of an asymptote to ensure that the eigenvalues are counted correctly.

Several modifications of the APW method were proposed prior to 1975, with the aim of circumventing these difficulties. Bross and co-workers [25, 26] proposed a modified APW method, in which multiple radial functions, chosen to have the same logarithmic derivative, are matched to the planewaves, with the requirement that both the value and first derivative of the wavefunction be continuous. This approach, which has strong similarities to the LAPW method, permits the determination of all the band energies with a single diagonalization. Koelling proposed an alternative, but related, method in which two radial functions are used, one having zero value on the sphere boundary and the other zero slope [85]. These functions, which are obtained by solving the radial equation subject to these boundary conditions, are matched to the planewaves to obtain continuity of the basis functions and their first derivatives. A potential difficulty with this approach is that, in general, imposition of the boundary conditions at the sphere radius results in basis functions that are mixtures of the regular and irregular solutions, *i.e.* they may not satisfy the physical boundary conditions at the nucleus. Nonetheless, Koelling has shown that this basis does improve the convergence for the d-bands of transition metals. Further, as in procedure of Bross, the asymptote problem does not occur. Other augmentations have been discussed by Marcus [114].

Andersen [2] (see also Ref. [86]) extended this work by proposing a method in which the basis functions and their derivatives are made continuous by matching to a radial function at fixed E_ℓ plus its derivative with respect to E_ℓ. This choice solved the problems with the APW method, discussed above, as well as providing a flexible and accurate band structure method. This is the LAPW method.

As may be expected, the first LAPW calculations [86] were within the MT approximation and used a model potential. However, shortly following this, fully self-consistent slab [78, 92, 68] and bulk [66] codes were developed, and general potential (no MT or other approximation to the charge density or potential, also called full-potential) calculations began appearing [66, 207, 76, 202, 203, 115, 17]. It was during this time that the power and accuracy of the method were demonstrated, largely through a series of calculations of surface and adsorbate electronic structures (see Wimmer *et al.* [208], for a review of this work). These and other demonstrations established the LAPW method as the method of choice for accurate electronic structure calculations for materials containing transition metal atoms.

4.2 The LAPW Basis and its Properties

In the LAPW method, the basis functions inside the spheres are linear combinations of radial functions, $u_\ell(r)Y_{\ell m}(\hat{\mathbf{r}})$ and their derivatives with respect to the linearization parameters, E_ℓ. The u_l are defined exactly as in the APW method (Eqn. 4.2), with a fixed E_ℓ. The energy derivative, $\dot{u}_\ell(r)Y_{\ell m}(\hat{\mathbf{r}})$ satisfies

$$\left[-\frac{\mathrm{d}^2}{\mathrm{d}r^2} + \frac{\ell(\ell+1)}{r^2} + V(r) - E_\ell\right] r\dot{u}_\ell(r) = ru_\ell(r), \tag{4.5}$$

in the non-relativistic case (see Chapter 5, for the relativistic case). These functions are matched to the values and derivatives of the planewaves on the sphere boundaries. Planewaves, augmented in this way, are the LAPW basis functions or LAPWs. In terms of this basis the wavefunctions are,

$$\varphi(\mathbf{r}) = \begin{cases} \Omega^{-1/2} \sum_{\mathbf{G}} c_{\mathbf{G}} e^{i(\mathbf{G}+\mathbf{k})\cdot\mathbf{r}} & \mathbf{r} \in I \\ \sum_{\ell m} [A_{\ell m} u_\ell(r) + B_{\ell m} \dot{u}_\ell(r) Y_{\ell m}(\hat{\mathbf{r}}) & \mathbf{r} \in S \end{cases}, \tag{4.6}$$

where the $B_{\ell m}$ are coefficients for the energy derivative, analogous to the $A_{\ell m}$. The LAPWs are just planewaves in the interstitial, as in the APW method. Inside the spheres the LAPWs have more variational freedom than APWs. This is because, if E_ℓ differs slightly from the band energy, ϵ, a linear combination, will reproduce the APW radial function constructed at the band energy.

$$u_\ell(\epsilon, r) = u_\ell(E_\ell, r) + (\epsilon - E_\ell)\dot{u}_\ell(\epsilon, r) + O((\epsilon - E_\ell)^2), \tag{4.7}$$

where $O((\epsilon - E_\ell)^2)$ denotes errors that are quadratic in this energy difference. For a converged (infinite) planewave set and a muffin-tin potential, the APW method yields exactly the correct wavefunction. In this case, going to the LAPW method introduces errors of order $(\epsilon - E_\ell)^2$ in the wavefunction; this, combined with the variational principle, yields errors of order $(\epsilon - E_\ell)^4$ in the band energy. Because of the high order of this error, the LAPWs form a good basis set over a relatively large energy region, so that all valence bands may typically be treated with a single set of E_ℓ. In the few instances for which this is not possible, the energy region of interest may be divided into a few (very rarely more than 2) windows and separate solutions carried out for each (see Chapter 5). This is an enormous simplification over the standard APW method. In the LAPW method accurate energy bands (at a given **k**-point) are obtained with a single diagonalization, while in the APW method one is needed for each band.

In general, if $u_\ell(R)$ is zero (*i.e.* at the asymptote), both its radial derivative and $\dot{u}_\ell(R)$ will be non-zero. Because of this, there is no asymptote problem in the LAPW method; the additional (continuous derivative) constraints ensure that the planewaves and local sectors do not decouple. Moreover, the LAPW basis has greater flexibility than the APW method inside the spheres, *i.e.* two radial functions instead of one. This means that there is no difficulty in treating non-spherical potentials inside the spheres; although the optimum value of E_ℓ is not known *a priori*, the flexibility arising from $\dot{u}_\ell$ allows an accurate solution. There is, however, a price to be paid for the additional flexibility of the LAPW basis. This arises from the requirement that the basis functions have continuous derivatives. Higher planewave cutoffs are required to achieve a given level of convergence. This requirement may be understood simply by considering a state that is highly localized inside the sphere. The wavefunction is composed of a linear combination of u_ℓ and $\dot{u}_\ell$ in the sphere and planewaves outside. The quality of the solution for such a state is determined by the wavefunction in the sphere. This means that the quality is determined by the ability of the linear combination of planewaves outside to match the ratio of the radial derivative of the wavefunction to its value. For a planewave, this ratio is at most $G = |\mathbf{G}|$. On the other hand, matching the value alone, as in the APW method, is much less demanding. For $\ell = 0$, this can be done with a single $G = 0$ planewave. The other limit is a nearly free electron band. In this case, the small high G Fourier components contribute little to the energy in an APW calculation, and accordingly the basis can be quite small. In the LAPW method, however, the derivatives bring down a factor G, and errors in the derivatives of the planewaves on the sphere produce errors in the value of the wavefunction inside.

Takeda and Kubler [191] proposed a generalization of the LAPW method in which N radial functions (each with its own energy parameter, $E_{\ell i}$) are matched to the value and $(N-1)$ derivatives of the interstitial planewaves. For $N = 2$, with $E_{\ell 1}$ near $E_{\ell 2}$ this is equivalent to the standard LAPW method, while for $N > 2$ residual errors due to the linearization can be removed. Unfortunately, the matching of higher derivatives makes this method much more slowly convergent with respect to the planewave cutoff than the standard LAPW method, and as a result it has not been widely used. Singh [166] modified this approach to avoid the matching of higher derivatives by adding specially constructed local orbitals to the basis to permit relaxation of the linearization without an increase in the planewave cutoff (see Chapter 5). This suggests the possibility of avoiding the slower convergence of the LAPW method relative to the APW method by using a similar construction.

This method, known as APW+LO, was implemented by Sjöstedt *et al.* [174], and shown to be highly effective in reducing the basis set sizes, especially for materials with large interstitial spaces and/or mixtures of atoms that require high planewave cut-offs with those requiring lower cut-offs. For example, in a supercell constructed to simulate a transition metal impurity in Al the standard LAPW method would require a rather large planewave cut-off, determined by the convergence properties for the transition metal atom. However, most of the unit cell consists of Al, which ordinarily requires a low cut-off. This is a result of the fact that planewaves are extended functions and so the global energy cut-off is determined by the hardest (in the sense of real space variation of the wavefunctions) region in the unit cell. In APW+LO implementations one can augment different atoms and angular momentum channels differently, so in particular the standard LAPW augmentation can be used for the Al atoms, and the APW+LO augmentation for the transition metal d states. This then lowers the global planewave cut-off needed. The number of planewaves, and therefore the size of the secular equation, scales as the cube of the planewave cut-off. Furthermore, the computational scaling of standard algorithms with the size of the secular equation is also cubic. As a result, the computational cost of a calculation scales nominally as the ninth power of the planewave cut-off, so the reduction in this cut-off enabled by the use of the APW+LO augmentation for the transition metal impurity results in a great savings in computational cost for the calculation.

Similarly, the cost of calculations can be greatly reduced by careful choice of the LAPW sphere radii. In the LAPW method, the spheres must be non-overlapping. This means that the sum of the sphere radii of two atoms must not exceed the distance between the two atoms. Thus the crystal structure defines the maximum pairwise sums of sphere radii, via the nearest neighbor distances, but it does not fix the choice of the individual radii. For example, in CsCl structure NiAl (lattice parameter 5.44 a_0) the nearest neighbor distance is 4.711 a_0, so the non-overlapping constraint on the sum of the sphere radii is, $r_{\text{Ni}} + r_{\text{Al}} \leq 4.711\ a_0$. Chemical and experimental considerations suggest the Al is effectively larger than Ni in this compound. For example, Ni on the Al site is a much more common defect in this material than Al atoms on Ni sites. However, it must be kept in mind that in the LAPW method the spheres are for computational convenience. They do not represent atomic sizes. Ni as a $3d$ transition metal would normally require a planewave cut-off $\sim 40\%$ higher than the simple sp metal, Al, for equal sphere radii and the same level of convergence. Since needed planewave cut-off often scales approximately as the inverse of the sphere radius, a much better starting point would be to select non-overlapping sphere radii in the ratio $r_{\text{Ni}}/r_{\text{Al}} \sim 1.4$. This would lead to a choice like $r_{\text{Ni}} = 2.70\ a_0$ and $r_{\text{Al}} = 2.00\ a_0$, *i.e.* contrary to naive chemical intuition for this material, but highly favorable from a computational point of view.

4.3 Role of the Linearization Energies

Before embarking on the more detailed discussion of the LAPW method and its implementation of the next chapter, it is worthwhile to remark further on the role of the linearization energies, E_ℓ; learning how to set these is a frequent source of grief for newcomers to the LAPW method.

It may seem initially straightforward to set the E_ℓ in the LAPW method. After all, the method derives from the APW method, and essentially reduces to it when E_ℓ is equal to the band energy, ϵ. Further, errors in the wavefunction (and charge density) are $O((\epsilon - E_\ell)^2)$ and the errors in the band energies are $O((\epsilon - E_\ell)^4)$. Thus it would seem that one needs simply to set the E_ℓ near the centers of the bands of interest to be assured of reasonable results, and one could in fact optimize the choice by computing the appropriate moments (depending on whether the goal is to minimize the error in ϵ or φ) of the density of states and using the known order of the errors to optimize the E_ℓ. Alternatively, one could envisage computing the total energy for several reasonable choices of E_ℓ and selecting the set that gave the lowest energy. Unfortunately, while these strategies work well in many cases, they fail miserably in many others. The reason for this failure is related to the presence of high lying and extended core states (also known as semi-core states) in many elements, particularly, the alkali metals, the rare earths, the early transition metals and the actinides.

As mentioned, the augmenting functions, $u_\ell Y_{\ell m}$ and $\dot{u}_\ell Y_{\ell m}$, are orthogonal to any core state that is strictly confined within the LAPW sphere. Unfortunately, this condition is never satisfied exactly except in the trivial case when there are no core states with the same ℓ. As a result, there can be a spurious component of the high lying core state contained in the valence wavefunctions as computed with the LAPW method. This problem does not arise in the APW method (with a MT potential and a converged planewave expansion, one may show that it cannot). The effects of this inexact orthogonality to extended core states in the LAPW method range from negligible to severe, and are sensitive to the choice of E_ℓ.

In the most severe case, there is sufficient overlap between the LAPWs and the core state that a badly converged core state appears in the spectrum (see Chapter 5, and also Ref. [115]). This spurious state (known as a ghost band) occurs above the true core state eigenvalue, and often in the valence part of the spectrum because the radial functions with E_ℓ are not well suited to representing the semi-core wavefunction. However, ghost bands are usually easily identified. They have very little dispersion, are unusually highly localized in the sphere, and have the ℓ character of the core state. Nonetheless, if present, they prevent the calculation from being carried through without modification of the E_ℓ. This can often preclude frozen phonon calculations (these require smaller sphere radii) in the early transition metals with the $\ell = 1$ energy parameter set to the center of the occupied valence bands. Two obvious options for eliminating ghost bands from the spectrum are (1) to set the offending energy parameter to the energy of the core state and (2) to raise the offending energy parameter to the next asymptote (where $u_\ell(R) = 0$).

Both of these options do eliminate the ghost state from the spectrum, but both are usually poor choices. They introduce subtle errors that can affect the results of a calculation, but may escape notice (*n.b.* certain properties particularly the lattice parameter and bulk modulus, which are often used as tests may

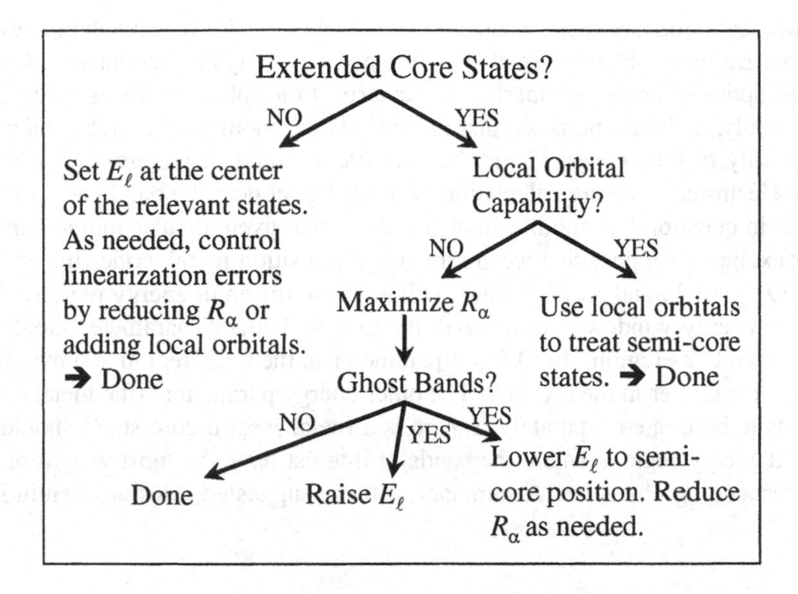

Figure 4.2. Procedure for setting the E_ℓ in the LAPW method. Note that the various ℓ components are to be considered separately.

be insensitive to the treatment of the valence p character in the early transition metals and their compounds, while other properties such as phonon frequencies and electric field gradients may be quite sensitive). Option (1) operates by providing more variational freedom for the ghost band, which then moves out of the valence spectrum to its proper place in the core. Unfortunately, this deprives the valence bands of variational freedom for that ℓ character. However, in the early transition metals (where the uppermost core p state yields the ghost band), there is significant $\ell = 1$ character in the valence bands near the zone boundary; this is lost when option (1) is used. With option (2) the ghost band is truly eliminated and without drastically reducing the variational freedom. Unfortunately, the valence states become increasingly non-orthogonal to the core state as the energy parameter is raised above it, and by the time E_ℓ reaches the asymptote in the early transition metals, a large part of the valence $\ell = 1$ character may be incorrectly described as core-like.

The ideal solution in such cases is to use the local orbital extension (see Chapter 5, for details); this allows an accurate treatment of both the core and valence states in a single energy window by adding extra variational freedom for selected ℓ. However, this option is not available in some codes. If this is the case, the best solution seems to be to set the sphere radius as large as possible, consistent with the structure, and if the ghost band still occurs, to move E_ℓ just high enough to move the ghost band safely above the Fermi energy, E_F.

However, in the rare cases when this requires moving E_ℓ very much (a few eV) above E_F the problems associated with option (2) may be encountered. In this case, option (1) may be superior, particularly if the sphere radius is reduced.

Finally, it should be noted that the various E_ℓ should be set independently. Typically, different energy bands have different orbital characters. For an accurate electronic structure calculation E_ℓ must be set near the band energy if the band in question has significant character of the given angular momentum, ℓ. For example, suppose that we are treating a transition metal oxide. In general, the O $2s$, O $2p$ and metal d bands will occur in different energy regions. If a single energy window is to be used, the O $\ell = 0$ energy parameter should be set in the O $2s$ region, the O $\ell = 1$ parameter in the O $2p$ region and the metal $\ell = 2$ parameter in the d bands. The other energy parameters (the metal $\ell = 1$ needs to be treated separately if there is a metal p semi-core state) should be set in energy regions where the bands of interest have the most weight of the corresponding character. A summary of these suggested rules is given in Fig. 4.2.

Chapter 5

NITTY-GRITTIES

5.1 Representations of the Charge Density and Potential

The efficiency of the LAPW basis derives from its use of carefully chosen representations of the wavefunctions in different regions. In particular, a spherical harmonics expansion on a radial mesh is used inside the spheres and a planewave expansion outside. With this choice, rapid variations of the wavefunctions inside the spheres pose no particular problems, and accordingly the method is well suited to all-electron calculations (*i.e.* no pseudopotential) as well as d- and f-electron materials. However, rapidly varying wavefunctions imply rapidly varying charge densities and potentials, and this requires that the representations of the charge density and potential be equally flexible.

The solution in the LAPW method is to use a dual representation for the charge and potential as well as the wavefunctions. A planewave expansion could be used in the interstitial and a spherical harmonic expansion inside. However, a direct implementation along these lines would yield an excessive number of parameters to be stored. (Note that the planewave cutoff for the charge density needs to be at least twice that of the wavefunctions, resulting in at least eight times as many coefficients.) Accordingly, symmetry is used to reduce the storage requirements; this has the added benefit of simplifying the construction of the charge density, and speeding up the synthesis of the Hamiltonian matrix.

The symmetries that are exploited are (1) inside a sphere the density has the site symmetry, (2) the interstitial density has the symmetry of the space group, (3) the density is a real quantity and (4) the densities within atoms related by a symmetry operation (equivalent atoms) are identical, apart from a rotation. This is done by using symmetry adapted expansions, stars in the interstitial and lattice harmonics within the inequivalent atoms.

Construction of the Stars

The symmetrized planewaves or stars, Φ_s, are defined by

$$\begin{aligned}\Phi_s &= \frac{1}{N_{op}} \sum_{\mathbf{R}} e^{i\mathbf{R}\,\mathbf{G}\cdot(\mathbf{r}-\mathbf{t}_{\mathbf{R}})} \\ &= \frac{1}{m_s} \sum_{m} \varphi_m e^{i\mathbf{R}_m\,\mathbf{G}\cdot\mathbf{r}}. \end{aligned} \tag{5.1}$$

Here the $\mathbf{R}$ are the rotational components of the space group operations, $\{\mathbf{R}|\mathbf{t}\}$, N_{op} is the number of space group operations and m_s is the number of independent planewaves in the star, which may be less than N_{op}. The phase factors, φ, defined by Eqn. 5.1, ensure that each star has the full symmetry of the lattice. It may be noted that (1) a given planewave occurs in only one star because of the group property, (2) for high symmetry lattices there are many fewer stars than planewaves, (3) all components of a star have the same $|\mathbf{G}|$, although not all planewaves with a given $|\mathbf{G}|$ need be in the same star, and most importantly (4) any scalar function that has the symmetry of the lattice can be expanded in stars. Further, the stars are orthogonal,

$$\frac{1}{\Omega} \int d^3\mathbf{r}\; \Phi_s^* \Phi_{s'} = \frac{1}{m_s} \delta_{s,s'}, \tag{5.2}$$

where Ω is the volume of the unit cell. The stars are constructed as shown in Fig. 5.1. A box that contains all planewaves up to the cut-off G_{max} is constructed in reciprocal space. All the reciprocal lattice vectors, $\mathbf{G}_i$ in the box are synthesized. if $|\mathbf{G}_i| \leq G_{max}$, then $\mathbf{G}_i$ is added to a list. After all the $\mathbf{G}_i$ in the box have been examined, the list is sorted by $|\mathbf{G}_i|$ (note that all elements of a star have the same length). The list is divided into sublists consisting of all planewaves of the same length and these are then further subdivided into lists of symmetry related planewaves (identified by applying the group operations to planewaves in the list). These form the stars. The phase factors are constructed using the space group operations,

$$\{\mathbf{R}|\mathbf{t}\}\mathbf{r} = \mathbf{R}\mathbf{r} + \mathbf{t}, \tag{5.3}$$

and (from Eqn. 5.1)

$$\varphi_m = \frac{m_s}{N_{op}} \sum_{\mathbf{R}\in\mathbf{m}} e^{-i\mathbf{R}\mathbf{G}\cdot\mathbf{t}}, \tag{5.4}$$

where the sum is over the space group operations that transform the representative $\mathbf{G}$ to the same $\mathbf{R}\mathbf{G}$. It can be shown that the number of such operations is the same for all $\mathbf{R}\mathbf{G}$ within a star.

For lattices that have inversion symmetry, the origin of the unit cell can be chosen on an inversion site, and in this case the phases are chosen so that the stars are real functions; as a result the star coefficients of the density and potential are also real. For lattices without inversion symmetry, this is not possible, because the star that contains **G** might not contain −**G**; in this case the star expansion coefficients are complex, although the reality of the density provides relationships among them. Alternatively, generalized stars consisting of real linear combinations of stars related by $\mathbf{G} \rightarrow -\mathbf{G}$ could be constructed, in principle. However, this would result in planewaves occurring in more than one star, and this would complicate certain operations required to construct the potential and symmetrize the charge density.

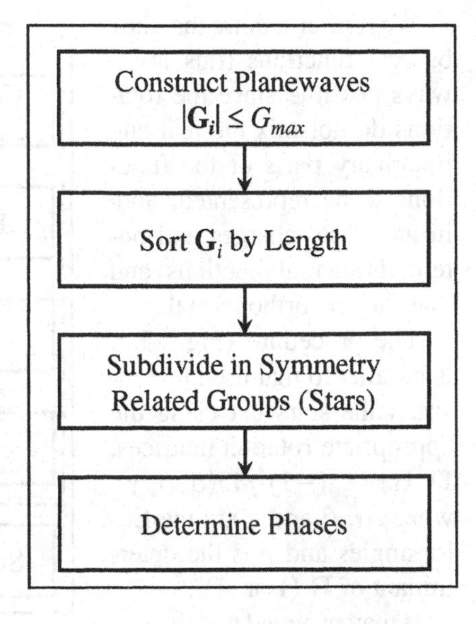

Figure 5.1. Construction of the stars.

The Lattice Harmonics

As mentioned, the lattice harmonics, $\mathbf{K}_\nu$, are the symmetrized spherical harmonics that are used for the sphere representations, just as the stars are used for the interstitial representations. However, unlike the stars, lattice harmonics are necessarily referenced to the center of the sphere in question. Because of this, the lattice harmonics are constructed using the site symmetry (operations that leave the atomic position invariant) rather than the perhaps higher space group symmetry.

$$\mathbf{K}_{\nu,\alpha}(\mathbf{r} - \mathbf{R}_\alpha) = \sum_m c^\alpha_{\nu,m} Y_{\ell m}(\mathbf{r} - \mathbf{R}_\alpha), \tag{5.5}$$

where $\mathbf{R}_\alpha$ is the position of the center of atom α. The sum is over m rather than ℓ and m because rotations of the $Y_{\ell m}$ do not couple different ℓ, just as rotations do not couple planewaves with different $|\mathbf{G}|$.

In order to determine the coefficients, $c_{\nu,m}$, we impose the requirement that the lattice harmonics be invariant under the rotations corresponding to the site symmetry.

We further require that they be real functions (this is always possible since the rotations do not mix the real and imaginary parts of the functions to be represented, and, in any case, charges and potentials are real functions) and that they be orthonormal.

The procedure (Fig. 5.2) is parallel to that used to construct the stars but using the appropriate rotation matrices, $\mathbf{D}(\mathbf{R}) \equiv (-1)^{\ell} p \mathbf{D}(\alpha, \beta, \gamma)$, where α, β and γ are the Euler angles and p is the determinant of $\mathbf{R}$ (1 or -1).

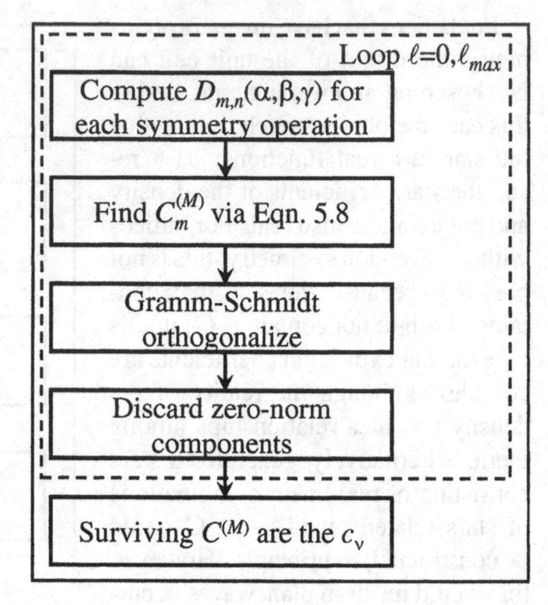

Figure 5.2. The construction of the lattice harmonics to represent the charge density and potential inside the LAPW spheres.

It may be noted that the $\ell = 0$ lattice harmonic is always present and has only one coefficient; the spherical component that it represents is typically handled separately. This is advantageous because the charge density and potential are nearly spherical close to the nucleus and therefore within some radius the $\ell \neq 0$ components can often be neglected.

The rotation matrices, $\mathbf{D}(\alpha, \beta, \gamma)$ are given by (see, for example, Brink and Satchler [23])

$$D_{m,m'}(\alpha, \beta, \gamma) = \mathrm{e}^{-im\alpha} d_{m,m'}(\beta)\, \mathrm{e}^{-im'\gamma}, \tag{5.6}$$

with

$$d_{m,m'}(\beta) = \sum_t (-1)^t \frac{[(\ell+m)!(\ell-m)!(\ell+m')!(\ell-m')!]^{1/2}}{(\ell+m-t)!(\ell-m'+t)!t!(t+m'-m)!} \\ \times cos^{2\ell+m-m'-2t}(\beta/2) sin^{2t+m'-m}(\beta/2), \tag{5.7}$$

where the sum on t is restricted to non-negative arguments of the factorials in the denominator.

In order to obtain real lattice harmonics, we apply the rotations to the real spherical harmonics, and sum over the operations $\mathbf{R}$, in the local site symmetry,

$$C_m^{(M)} = \begin{cases} \sum_{\mathbf{R}} [D_{m,M}(\mathbf{R}) + (-1)^M D_{m,-M}(\mathbf{R})] & M \geq 0 \\ \sum_{\mathbf{R}} i[D_{m,M}(\mathbf{R}) - (-1)^M D_{m,-M}(\mathbf{R})] & M < 0, \end{cases} \tag{5.8}$$

At this point the $C^{(M)}$ are Gramm-Schmidt orthonormalized, and those that have zero norm or are linearly dependent are discarded. The remaining $C^{(M)}$ are exactly the c_ν, where ν is just a sequential numbering of the survivors.

As mentioned, the density and potential inside each sphere are expanded in lattice harmonics on a discrete radial mesh, r_i (note that the same radial mesh is used for the wavefunctions). An accurate representation requires both a sufficient number of lattice harmonics and a sufficiently dense radial mesh. A common choice for the latter is a logarithmic mesh,

$$r_{i+1} = r_i e^{\delta_x}, \tag{5.9}$$

where the last mesh point, $r_{i=i_{max}} = R_\alpha$. A high degree of convergence is typically attained with $\delta_x \sim 0.03$, provided that proper algorithms are used for integrations and for solving differential equations. In fact, the use of logarithmic meshes is particularly convenient for integrations and numerical solutions of differential equations (*e.g.* the radial Schrodinger's equation) because if $r = e^x$, then $dr = r dx$, which results in relatively simple expressions.

Finally, since the charge density and potentials inside symmetry related spheres are the same (modulo a rotation), only the expansion within the representative atoms need be retained (*e.g.* a single atom in the two atom per unit cell diamond structure). This means that either the lattice harmonics must be explicitly rotated when using the density or potential in the non-representative atoms, or, as is done in practice, it is understood that the $Y_{\ell m}$ are in a coordinate system that is rotated by the symmetry operation that relates the representative atom to the equivalent atom in question.

5.2 Solution of Poisson's Equation

The potential to be used in the KS equations consists of an exchange correlation term (discussed in the next section) and a Coulomb term, $V_C(\mathbf{r})$, which is the sum of the Hartree potential, $V_H(\mathbf{r})$ and the nuclear potential. $V_C(\mathbf{r})$ is determined by the charge (electronic plus nuclear) via Poisson's equation (in atomic units, $e^2 = 1$),

$$\nabla^2 V_C(\mathbf{r}) = 4\pi\rho(\mathbf{r}). \tag{5.10}$$

Given boundary conditions, integration of this equation can be performed efficiently in a small region. However, in general the solution in real space is not at all straightforward. On the other hand, Poisson's equation is diagonal in reciprocal space, making the solution trivial in principle.

$$V_C(\mathbf{G}) = \frac{4\pi\rho(\mathbf{G})}{|\mathbf{G}|^2}. \tag{5.11}$$

Unfortunately, in the LAPW method $\rho(\mathbf{r})$ contains the rapidly varying core density as well as the delta function-like nuclear charge density, and as a result the Fourier expansion, $\rho(\mathbf{G})$ is not convergent. Thus the short range behavior of density in the atomic cores complicates a reciprocal space formulation, while the long range character of the Coulomb potential complicates real space approaches. Hamann [66] and Weinert [204] have developed a hybrid method that circumvents these problems. The technique is based on three observations: (1) the interstitial charge density is smooth, with the rapidly varying component being confined within the spheres; (2) the Coulomb potential outside a sphere depends only on the charge outside the sphere and the multipoles of the charge inside, and (3) planewaves are an overcomplete description of the interstitial charge density, since the Fourier transform of any function that is confined within the spheres may be added, without changing the interstitial charge.

The procedure, which is known as the pseudo-charge method, proceeds as illustrated by Fig. 5.3. The multipoles of the planewave expansion of the interstitial charge density are calculated within each sphere, as are the multipoles, $q_{\ell m}$, of the true charge density. The latter may be determined directly from integrals of the radial parts of the lattice harmonic expansion, ρ_ν.

$$q_{\ell m} = \sum_\nu C_{\nu,m} \int_0^{R_\alpha} r^{\ell+2} \rho_\nu(r) \delta_{\ell,\ell_\nu} \, \mathrm{d}r, \tag{5.12}$$

where r is the radial coordinate $|\mathbf{r} - \mathbf{r}_\alpha|$, $\mathbf{r}_\alpha$ being the location of the sphere in question and R_α is the corresponding sphere radius. The planewave multipoles, $q_{\ell m}^{PW}$ are calculated from the Bessel function expansion of the planewaves,

$$\mathrm{e}^{i\mathbf{G}\cdot\mathbf{r}} = 4\pi \mathrm{e}^{i\mathbf{G}\cdot\mathbf{r}_\alpha} \sum_{\ell m} i^\ell j_\ell(|\mathbf{G}||\mathbf{r} - \mathbf{r}_\alpha|) Y_{\ell m}^*(\mathbf{G}) Y_{\ell m}(\mathbf{r} - \mathbf{r}_\alpha). \tag{5.13}$$

The required integral is

$$\int_0^R r^{\ell+2} j_\ell(Gr) \, \mathrm{d}r = \begin{cases} \frac{R^{\ell+3} j_\ell(GR)}{GR} & G \neq 0 \\ \frac{R^3}{3}\delta_{\ell,0} & G = 0. \end{cases} \tag{5.14}$$

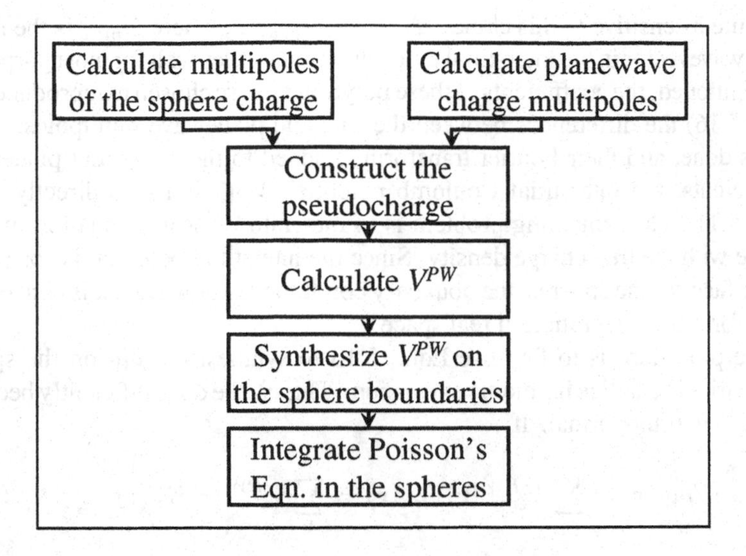

Figure 5.3. Pseudocharge method for solving the Poisson equation.

Next a pseudo-charge is constructed, such that the pseudo-charge equals the original charge in the physically relevant interstitial region, but has the same multipoles within each sphere as the true sphere charge density. This is accomplished by adding smooth functions that are zero outside the spheres and have multipoles equal to the differences between the sphere charge multipoles and the original planewave multipoles. The choice of these functions is arbitrary, provided that the Fourier transforms can be calculated readily. A convenient choice is to use a polynomial form,

$$\tilde{\rho}_\alpha(\mathbf{r}) = \sum_{\ell m} Q_{\ell m} \frac{1}{R_\alpha^{\ell+3}} (\frac{r}{R_\alpha})^\ell (1 - \frac{r^2}{R_\alpha^2})^N Y_{\ell m}(\hat{\mathbf{r}}), \tag{5.15}$$

where r is now with respect to the center of the sphere. This functional form has $(N-1)$ continuous derivatives and an analytic Fourier transform (use Eqn. 5.13 for the planewave to obtain products of Bessel functions and polynomials). The multipole moments, $\tilde{q}_{\ell m}$, are

$$q_{\ell m} = Q_{\ell m} \frac{\Gamma(\ell + 3/2)\Gamma(N+1)}{2\Gamma(\ell + N + 5/2)} = Q_{\ell m} \frac{2^N N!(2\ell+1)!!}{(2\ell + 2N + 3)!!}. \tag{5.16}$$

Weinert [204] discussed the convergence of the Fourier transform as a function of N, and presented a table giving optimum values. In practice, the results are quite insensitive to this choice; $N \approx R_\alpha G_{max}/2$, where G_{max} is the maximum wavevector in the interstitial representation, serves well for most purposes. As mentioned, the coefficients of these polynomials are chosen to reproduce (via Eqn. 5.16) the differences between the true and planewave multipoles. Once this is done, and their Fourier transform is added to the interstitial planewave coefficients, the interstitial Coulomb potential, V_{PW} is found directly using Eqn. 5.11. The remaining problem is to integrate Poisson's equation in each sphere with the true charge density. Since the interstitial potential is correct on the surfaces of the spheres, the boundary condition is known and it is convenient to perform the integration in real space.

The procedure is to first use Eqn. 5.13 to synthesize V_{PW} on the sphere boundaries in a lattice harmonic expansion. This can be done efficiently because the $\mathbf{K}_\nu$ are orthonormal. If

$$V_{PW}(\mathbf{r}) = \sum_{\ell m} V_{\ell m}^{PW}(r) Y_{\ell m}(\hat{\mathbf{r}}) = \sum_{\nu} V_{\nu}^{PW}(r) \mathbf{K}_\nu(\hat{\mathbf{r}}), \tag{5.17}$$

then

$$V_{\nu}^{PW}(r) = \sum_{m} c_{\nu,m} V_{\ell m}^{PW}(r), \tag{5.18}$$

where ℓ corresponds to that of the given $\mathbf{K}_\nu$. Next a standard Green's function approach is used in real space to compute the potential in the sphere.

$$V_\nu(r) = V_{\ell m}^{PW}(R)[\frac{r}{R}]^\ell + \frac{4\pi}{2\ell+1}[\frac{1}{r^{\ell+1}} \int_0^r (r')^{\ell+2} \rho_\nu(r') \mathrm{d}r' + r^\ell \int_r^R (r')^{1-\ell} \rho_\nu(r') \mathrm{d}r' - \frac{r^\ell}{R^{2\ell+1}} \int_0^R (r')^{\ell+2} \rho_\nu(r') \mathrm{d}r'], \tag{5.19}$$

where R denotes the sphere radius, $\rho_\nu(r)$ are the radial parts of the lattice harmonic expansion of the charge density, and the potential is in atomic units. For $\ell = 0$, the nuclear charge is understood to be included in ρ_0. This amounts to adding the nuclear contribution to the Coulomb potential.

5.3 The Exchange Correlation Potential

The LDA exchange-correlation potential, unlike the Coulomb potential, is nonlinear. Because of this, it must be calculated in real space, where it is fortunately diagonal. The problem then amounts to transforming the charge density to a real space representation, calculating the exchange-correlation potential, $V_{xc}(r)$ and back transforming to the LAPW representation. The procedure is illustrated in Fig. 5.4.

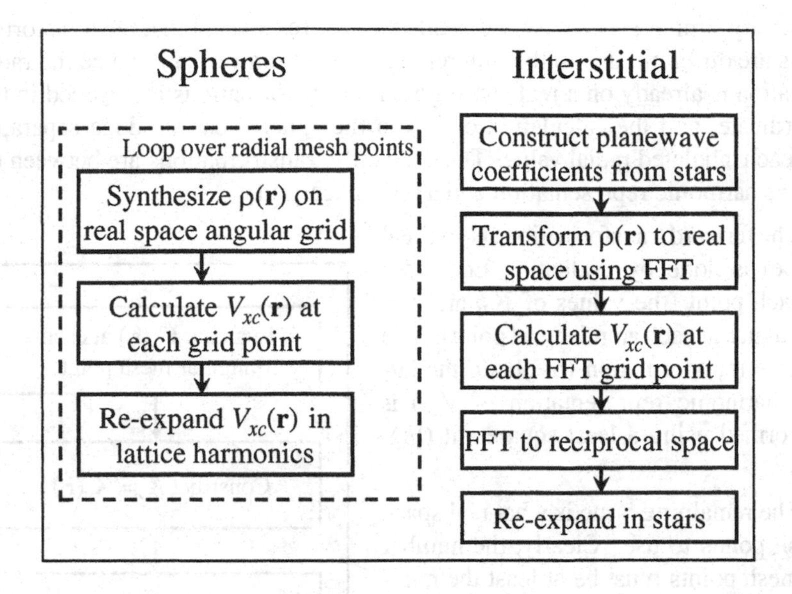

Figure 5.4. The exchange-correlation potential.

The generalization to GGA exchange-correlation functionals is conceptually straight-forward; the density and local gradients and second derivatives needed to evaluate the GGA V_{xc} are generated in real space and used in the same way as for the LDA. The generalization to collinear spin-polarized calculations consists in transforming the two spin densities to real space, calculating the two components of $V_{xc}(r)$ and then back transforming to the LAPW representation.

The modification for non-collinear magnetic systems is to use a local spin rotation to transform locally at each real space point, r, to a collinear case, calculate the spin polarized $V_{xc}(r)$, and then rotate back. The result is that the direction of $V_{xc}(r)$ will be the same as that of the magnetization at each r and therefore that V_{xc} will not exert any local torques on the magnetization (this is an approximation, as the true exchange-correlation function would produce torques via the correlation part of the kinetic energy; this already arises in principle at the GGA level, as discussed in Chapter 2).

The real space representation of the interstitial charge is obtained directly, via the Fourier transform. The planewave coefficients are constructed from the star representation of the interstitial charge using Eqn. 5.1. A FFT is then used to transform to values on the real space grid. $V_{xc}(\mathbf{r})$ is calculated on each mesh point (including those inside the spheres to avoid Gibb's oscillations between the mesh points). A FFT is then used to transform V_{xc} back to a planewave representation, from which star coefficients are obtained.

A very similar scheme is used inside the spheres, except that the transformations are different, due to the different representation of $\rho(\mathbf{r})$. Since the radial variation is already on a real space grid, no transformations are needed in this coordinate, and the calculation of V_{xc} in the spheres can be done separately for each tabulated radial value. Thus, the only transformations are between the lattice harmonic representation and a real space grid.

The forward transformation ($\mathbf{K}_\nu$ to real space) is done by evaluating Eqn. 5.5 at each point (the values of $\mathbf{K}_\nu$ are precalculated at each angular grid point). The reverse transformation (to obtain the lattice harmonic representation of V_{xc}) is performed using a least squares fit (Fig. 5.5).

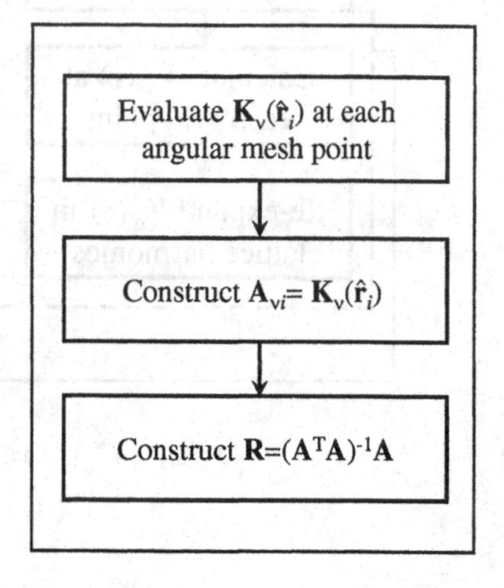

Figure 5.5. $\mathbf{A}^T$ and $\mathbf{R}$ (which can be precalculated outside the radial loop) are used to transform from lattice harmonics to real space and vice-versa.

The remaining issue is what real space mesh points to use. Clearly, the number of mesh points must be at least the number of lattice harmonics (including $\ell = 0$). Otherwise, the least squares fit will be under determined. In addition, since V_{xc} is non-linear, it will contain lattice harmonics with higher angular momenta than the input charge density. Although these are to be discarded, they can introduce errors in the fit. Therefore it is desirable to use more than the minimum number of points. Mattheiss and Hamann [115] used a 72 point mesh that integrates all $Y_{\ell m}$ for $\ell \leq 14$ exactly [117]. An alternate but less elegant choice (needed if higher ℓ lattice harmonics are used) is to scatter a large number of points on the surface of the sphere. These may then be symmetrized using the site symmetry to avoid introducing a spurious symmetry breaking that could otherwise arise via the fits involved in calculating the exchange-correlation potential.

5.4 Synthesis of the LAPW Basis Functions

As discussed in Chapter 4, the LAPW basis functions are planewaves in the interstitial. These are matched onto numerical radial functions inside the spheres, with the requirement that the basis functions and their first derivatives are continuous at the boundary. Thus synthesis of the LAPW basis functions amounts to determining (1) the radial functions, $u_\ell(r)$ and $\dot{u}_\ell(r)$ and (2) the coefficients $a_{\ell m}$ and $b_{\ell m}$ that satisfy the boundary condition.

The boundary conditions also provide a simple prescription for determining reasonable angular momenta cutoffs, ℓ_{max} for the sphere representation in terms of the planewave cutoff, K_{max}. A sensible strategy is to tune these cutoffs so that the two truncations will be well matched, thereby avoiding Gibbs-like oscillations. This may be done by noting that a given ℓ_{max} allows representation of functions with a maximum $2\ell_{max}$ nodes along a great circle around the sphere, *i.e.* a distance of $2\pi R_\alpha$, or $\ell_{max}/(\pi R_\alpha)$ nodes/a_0 in atomic units. On the other hand, K_{max} corresponds to a planewave with K_{max}/π nodes/a_0. Equating these suggests a criterion, $R_\alpha K_{max} = \ell_{max}$, and, in fact, this works well in practice. Since LAPW calculations are usually adequately well converged for $R_\alpha K_{max}$ in the range 7-9; this results in $\ell_{max} \sim 8$ as a typical value.

Construction of the Radial Functions

In a non-relativistic implementation, the radial functions, $u_\ell(r)$ are the solutions of the radial Schrodinger equation with the spherically averaged crystal potential, at the linearization energy E_ℓ. As discussed in Chapter 4, in atomic (Rydberg) units we have

$$[-\frac{\mathrm{d}^2}{\mathrm{d}r^2} + \frac{\ell(\ell+1)}{r^2} + V(r) - E_\ell]ru_\ell(r) = 0, \tag{5.20}$$

where $V(r)$ is the radial potential from the $\ell = 0$ lattice harmonic (*n.b.* depending on the normalization of this lattice harmonic, there may be a factor of $(4\pi)^{-1/2}$) and the boundary condition $ru_\ell(0)$ is to be enforced. Taking the derivative with respect to the linearization energy, one obtains,

$$[-\frac{\mathrm{d}^2}{\mathrm{d}r^2} + \frac{\ell(\ell+1)}{r^2} + V(r) - E_\ell]r\dot{u}_\ell(r) = ru_\ell(r). \tag{5.21}$$

These differential equations may be solved on the radial mesh using standard, *e.g.* predictor-corrector, methods (a good reference is the book by Press *et al.* [150]). However, since Eqn. 5.20 is linear, the norm of u_ℓ is undetermined, and furthermore, given a solution, $\dot{u}_\ell$ of Eqn. 5.21, $\dot{u}_\ell + cu_\ell$ with arbitrary c is also a solution. In practice, it is convenient to enforce the normalization,

$$\int_0^{r_\alpha} [ru_\ell(r)]^2 \mathrm{d}r = 1, \tag{5.22}$$

and to orthogonalize u_ℓ and $\dot{u}_\ell$

$$\int_0^{r_\alpha} r^2 u_\ell(r)\dot{u}_\ell(r)\mathrm{d}r = 0, \tag{5.23}$$

With this choice, the norm of $\dot{u}_\ell$ ($||\dot{u}_\ell||$) provides an indication as to the range over which the energy linearization is a reasonable approximation. In particular, the linearization errors are acceptable for most purposes when $||\dot{u}_\ell|| \times |E_\ell - \varepsilon| \leq 1$, where E_ℓ is/are the energy parameter(s) for those ℓ for which the band in question has significant character and ε is the band energy (in a total energy calculation, for example, the criterion should be satisfied for all occupied states (see Chapter 4, and the paper of Andersen [2], for more discussion of linearization errors).

Several options are available if a choice of E_ℓ that satisfies the above criterion cannot be determined. These are (1) to divide the energy range into windows (also known as panels) and treat each window separately with E_ℓ appropriate for the states in it, (2) to relax the linearization using a local orbital extension (this is effectively a quadratic method), or (3) to reduce the sphere sizes, thereby reducing $||\dot{u}_\ell||$. The first two options (discussed in detail in later in this chapter) are commonly used. The latter, while generally applicable, results in substantial increases in the basis set size, and these increases are often computationally prohibitive and further may exacerbate problems due to high lying extended core states (*e.g.* ghost bands), as discussed earlier. However, iterative implementations of the LAPW method impose smaller penalties for increased basis set size, and in these methods (3) may be a reasonable alternative [55].

Relativistic Radial Functions

Relativistic corrections are important only when the kinetic energy is large. Since the band energies of interest in solids are small, this means that relativistic effects need only be incorporated in regions where the potential is strongly negative, *i.e.* near the nuclei. In the LAPW method, this means that relativistic effects can be safely neglected in the interstitial region, and the only modifications are to the radial functions in the spheres and the components of the Hamiltonian that operate on them.

The relativistic modification is, therefore, to replace Eqns. 5.20 and 5.21 by the corresponding Dirac equation and its energy derivative, and retain the relativistic terms when evaluating the sphere contribution to the Hamiltonian matrix elements. It is, however, convenient to neglect spin-orbit effects at this stage (the scalar relativistic approximation), since otherwise the size of the secular equation is doubled. If important, spin-orbit effects can be taken account of latter using the low lying band wavefunctions as the basis for a second variational step. The exception is for calculations with non-collinear magnetism, in which case the secular equation is doubled regardless. Koelling and Harmon [87] (see also Rosicky *et al.* [154] Wood and Boring [209], Takeda [190], and MacDonald *et al.* [113]) have presented a technique for solving the Dirac equation in a spherically symmetric potential, in which spin-orbit effects are initially neglected, but may be incorporated afterward.

The solution of the Dirac equation is conventionally written as,

$$\Phi_{\kappa\mu} = \begin{bmatrix} g_\kappa \chi_{\kappa\mu} \\ -i f_\kappa \sigma_r \chi_{\kappa\mu} \end{bmatrix}, \tag{5.24}$$

where κ is the relativistic quantum number, $\chi_{\kappa\mu}$ is a two-component spinor and the radial coordinate has been suppressed.

Koelling and Harmon define a new function,

$$\phi_\kappa = \frac{1}{2Mc} g'_\kappa. \tag{5.25}$$

Here the prime denotes the radial derivative, m is the mass, c is the speed of light and

$$M = m + \frac{1}{2c^2}(E - V), \tag{5.26}$$

at energy, E. Dropping the spin-orbit term, the solution is rewritten with the usual non-relativistic ℓm quantum numbers as

$$\Phi_{\ell ms} = \begin{bmatrix} g_\ell Y_{\ell m} \chi_s \\ \frac{i}{2Mc}\sigma_r(-g'_\ell + \frac{1}{r} g_\ell \sigma \cdot L) Y_{\ell m} \chi_s \end{bmatrix}, \tag{5.27}$$

where χ_s is the usual non-relativistic spinor. Defining $P_\ell = r g_\ell$ and $Q_\ell = r c g_\ell$, the scalar relativistic equations become

$$P'_\ell = 2MQ_\ell + \frac{1}{r}P_\ell, \tag{5.28}$$

and

$$Q'_\ell = -\frac{1}{r}Q_\ell + [\frac{\ell(\ell+1)}{2Mr^2} + (V - E_\ell)]P_\ell. \tag{5.29}$$

These can be solved numerically in the same way as the non-relativistic Schrodinger equation (*e.g.* standard predictor-corrector), given the boundary condition,

$$\lim_{r \to 0} \frac{Q}{P} = c\frac{[\ell(\ell+1) + 1 - (2Z/c)^2]^{1/2} - 1}{(2Z/c)}. \tag{5.30}$$

The spin-orbit term (*n.b.* there are higher order terms in M^{-1} that have been neglected) may be included by adding $-(V'/4M^2c^2)(\kappa + 1)P$ to the right hand side of Eqn. 5.29. The energy derivatives, needed for the linearization, are analogous to the non-relativistic case.

$$\dot{P}'_\ell = 2(\dot{M}Q_\ell + M\dot{Q}_\ell) + \frac{1}{r}\dot{P}_\ell, \tag{5.31}$$

and

$$\dot{Q}'_\ell = -\frac{1}{r}\dot{Q}_\ell + [\frac{\ell(\ell+1)}{2Mr^2} + (V - E_\ell)]\dot{P}_\ell - [\frac{\ell(\ell+1)\dot{M}}{2M^2r^2} + 1]P_\ell. \tag{5.32}$$

From the solution, P_ℓ and Q_ℓ, the large and small components, g_ℓ and f_ℓ can be determined via the definitions of P_ℓ, Q_ℓ and ϕ_ℓ. Both the large and small components are to be used in constructing the charge density or evaluating matrix elements (*e.g.* for the non-spherical component of the Hamiltonian). Thus the quantity replacing u^2 in the normalization (Eqn. 5.22) is g^2+f^2. However, at the sphere boundary the small component is assumed to vanish consistent with the non-relativistic treatment in the interstitial. Therefore, in enforcing the matching at the boundary, only the large component and its derivative is used.

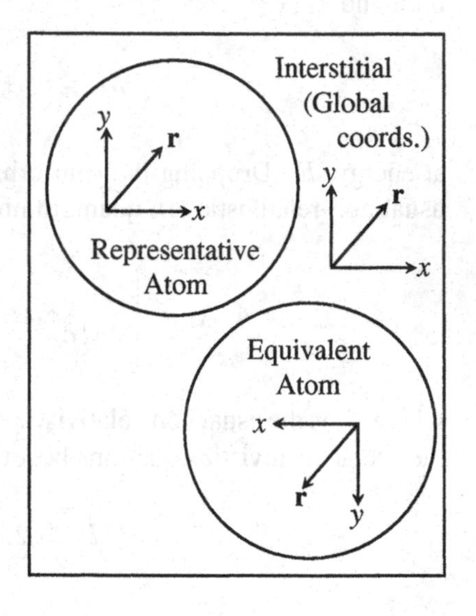

Figure 5.6. The rotated coordinate system inside equivalent (*i.e.* symmetry related) atoms. Note that the representative atom is in the global frame.

The $a_{\ell m}$ and $b_{\ell m}$

As mentioned, the LAPW basis functions are constructed to be continuous up to first derivatives across the sphere boundaries. This determines the coefficients $a_{\ell m}(\mathbf{k}+\mathbf{G})$ and $b_{\ell m}(\mathbf{k}+\mathbf{G})$ for each planewave and atom (*n.b.* the atom index, α, is suppressed). In order to do this two things are used: (1) the value and radial derivative of the angular momentum decomposition of the planewaves (Eqn. 5.13) and (2) the value and radial derivative of the u_ℓ and $\dot{u}_\ell$, both at the sphere radius. Continuity of the angular derivatives arises from the continuity of each ℓm component; first derivatives of products of radial functions and $Y_{\ell m}$ change ℓ by one at most.

Thus continuity of the ℓm components up to $\ell = \ell_{max}$ ensures that the angular derivative is continuous up to $\ell_{max} - 1$.

In cases where there are symmetry related atoms (*e.g.* the two equivalent C atoms in the diamond unit cell), it is convenient not to use the global coordinate system for the $Y_{\ell m}$. Rather, inside a given sphere, the coordinates are rotated about the center using the rotational part, $\mathbf{R}$ (see Eqn. 5.3), of the space group operation that generates the atom in question from the representative atom (Fig. 5.6). Since the lattice harmonic representation is also rotated, this choice simplifies the construction and symmetrization of the charge density and the operation of the non-spherical component of the Hamiltonian in the spheres.

Thus we obtain

$$\frac{1}{\Omega^{1/2}} e^{i(\mathbf{k}+\mathbf{G})\cdot\mathbf{r}} = \frac{4\pi i^{\ell}}{\Omega^{1/2}} e^{i(\mathbf{k}+\mathbf{G})\cdot\mathbf{r}_{\alpha}} \times \sum_{\ell m} j_{\ell}(|\mathbf{k}+\mathbf{G}||\mathbf{r}-\mathbf{r}_{\alpha}|) Y^{*}_{\ell m}(\mathbf{R}(\mathbf{k}+\mathbf{G})) Y_{\ell m}(\mathbf{R}(\mathbf{r}-\mathbf{r}_{\alpha})), \tag{5.33}$$

which is rewritten in terms of a structure factor $F_{\ell m,\alpha}(\mathbf{k}+\mathbf{G})$

$$\frac{1}{\Omega^{1/2}} e^{i(\mathbf{k}+\mathbf{G})\cdot\mathbf{r}} = \frac{1}{R_{\alpha}^{2}} \sum_{\ell m} F_{\ell m,\alpha}(\mathbf{k}+\mathbf{G}) j_{\ell}(|\mathbf{k}+\mathbf{G}||\mathbf{r}-\mathbf{r}_{\alpha}|) Y_{\ell m}(\mathbf{R}(\mathbf{r}-\mathbf{r}_{\alpha})). \tag{5.34}$$

Here $\mathbf{r}_{\alpha}$ is the location of the sphere, $\mathbf{r}$ is the point in question (in the global frame), R_{α} is the sphere radius, and $F_{\ell m,\alpha}(\mathbf{k}+\mathbf{G})$ is defined by Eqns. 5.33 and 5.34. In this last equation, all the radial dependence is explicitly contained in the argument of the Bessel function, and so the radial derivative is given by a similar expression, but containing j'_{ℓ} instead of j_{ℓ}.

We recall that the basis functions for $\mathbf{r}$ inside the sphere, α, are given by

$$\phi_{\mathbf{k}+\mathbf{G}}(\mathbf{r}) = \sum_{\ell m} Y_{\ell m}(\mathbf{R}(\mathbf{r}-\mathbf{r}_{\alpha}))[a_{\ell m} u_{\ell}(|\mathbf{r}-\mathbf{r}_{\alpha}|) + b_{\ell m} \dot{u}_{\ell}(|\mathbf{r}-\mathbf{r}_{\alpha}|)]. \tag{5.35}$$

So, using the matching conditions to determine $a_{\ell m}$ and $b_{\ell m}$ amounts to solving a 2×2 linear system. It is normally most efficient to calculate the $F_{\ell m,\alpha}$ and the $a_{\ell m}/F_{\ell m,\alpha}$ and $b_{\ell m}/F_{\ell m,\alpha}$ (a 2×2 real linear system) separately and then form the product.

5.5 Synthesis of the Hamiltonian and Overlap Matrices

The matrix elements, $S_{\mathbf{GG}'}$ and $H_{\mathbf{GG}'}$, of the overlap and Hamiltonian, respectively, are

$$S_{\mathbf{GG}'} = <\phi_{\mathbf{G}}|\phi_{\mathbf{G}'}>, \tag{5.36}$$

and

$$H_{\mathbf{GG}'} = <\phi_{\mathbf{G}}|H|\phi_{\mathbf{G}'}>. \tag{5.37}$$

These, like the LAPW basis functions, are **k** dependent, but for compactness, the **k** index has been suppressed. $S_{\mathbf{GG}'}$ and $H_{\mathbf{GG}'}$ may be decomposed into interstitial and sphere components, and the latter further decomposed into spherical (kinetic energy plus $\ell = 0$ part of the potential) and non-spherical terms in the case of the Hamiltonian.

$$S_{\mathbf{GG}'} = \frac{1}{\Omega}\int_{\Omega} \mathrm{d}^3\mathbf{r}\, \mathrm{e}^{i(\mathbf{G}'-\mathbf{G})\cdot\mathbf{r}}\, \Theta(\mathbf{r}) + \sum_{\alpha} S_{\alpha}(\mathbf{G},\mathbf{G}'), \tag{5.38}$$

and

$$H_{\mathbf{GG}'} = \frac{1}{\Omega}\int_{\Omega} \mathrm{d}^3\mathbf{r}\, \Theta(\mathbf{r})\mathrm{e}^{-i(\mathbf{G}+\mathbf{k})\cdot\mathbf{r}}[T + V_{PW}(r)]\, \mathrm{e}^{i(\mathbf{G}'+\mathbf{k})\cdot\mathbf{r}} + \sum_{\alpha}[H_{\alpha}(\mathbf{G},\mathbf{G}') + V_{\alpha}^{NS}(\mathbf{G},\mathbf{G}')], \tag{5.39}$$

where the integrals are over the unit cell, which has volume Ω. T is the kinetic energy operator, the $S_{\alpha}(\mathbf{G},\mathbf{G}')$ are the contributions to the overlap from sphere, α, the $H_{\alpha}(\mathbf{G},\mathbf{G}')$ are the corresponding spherical contributions to the Hamiltonian, the $V_{\alpha}^{NS}(\mathbf{G},\mathbf{G}')$ are the contributions from the $\ell \neq 0$ potential in the sphere, and $\Theta(\mathbf{r})$ is a step function defined to be zero inside any sphere and unity in the interstitial.

The LAPW basis functions transform as planewaves. As such, it is easy to exploit inversion symmetry, if present, by choosing the origin of the unit cell to coincide with an inversion center. With this choice, the Hamiltonian and overlap become real symmetric matrices, making their construction and diagonalization more efficient than for the complex Hermetian case that arises otherwise.

The Interstitial Contribution

In the absence of the step function, $\Theta(\mathbf{r})$, computation of the interstitial contribution would be straightforward; the overlap would be $\delta_{\mathbf{G},\mathbf{G}'}$, while $V_{PW}(\mathbf{r})$ is a local potential, diagonal in real space, and T would be diagonal in **G** space. Thus the key to evaluating the interstitial component is an understanding of the operation of the step function, $\Theta(\mathbf{r})$.

Initially, it may seem that since $\Theta(\mathbf{r})$ is diagonal in real space, it can be operated by multiplying on the discrete real space (FFT) mesh. This is, however, incorrect. The Fourier transform of $\Theta(\mathbf{r})$ is non-convergent, and this leads to errors when a discrete real space mesh is used (this is equivalent to a cut-off in momentum space). To see this, consider operating $\Theta(\mathbf{r})$ on a constant function $f(\mathbf{r}) = \Omega^{-1}$. The matrix element in this case is the interstitial volume fraction. However, the result obtained by direct multiplication with $\Theta(\mathbf{r})$ is the fraction of the discrete mesh points that are in the interstitial; this will go to the correct result in the limiting case of an infinitely dense mesh, but is not in general exact for any finite mesh and is very slowly convergent (as N^{-1}, where N is the number of mesh points). Thus an understanding of the function upon which $\Theta(\mathbf{r})$ is to be operated is needed. In particular, the fact that the function that we need to integrate with consists of a finite Fourier series is used. For any function, $f(\mathbf{r})$ having a finite Fourier series with maximum $|\mathbf{G}| = G_{max}$,

$$\begin{aligned}\frac{1}{\Omega}\int_{\Omega} \mathrm{d}^3\mathbf{r}\, f(\mathbf{r})\Theta(\mathbf{r}) &= \sum_{|\mathbf{G}|\le G_{max}} f(\mathbf{G})\Theta(-\mathbf{G}) \\ &= \frac{1}{\Omega}\int_{\Omega} \mathrm{d}^3\mathbf{r}\, f(\mathbf{r})\tilde{\Theta}(\mathbf{r})\,, \end{aligned} \tag{5.40}$$

provided that $\tilde{\Theta}(\mathbf{G}) = \Theta(\mathbf{G})$ for all $|\mathbf{G}| \le G_{max}$ (note the factors of Ω^{-1}, which depend on the convention for the normalization of the FFT). Thus, because of the smoothness of the functions that multiply $\Theta(\mathbf{r})$ in the integrands, we are free to chose a smooth $\tilde{\Theta}(\mathbf{r})$, subject to the condition above, and use it in place of $\Theta(\mathbf{r})$, without introducing any approximation. The most convenient choice is to construct $\Theta(\mathbf{G})$ analytically and truncate it at G_{max}.

$$\tilde{\Theta}(\mathbf{G}) = \begin{cases} \delta_{\mathbf{G},0} - \sum_\alpha \frac{4\pi R_\alpha^3}{\Omega}\frac{j_1(|\mathbf{G}|R_\alpha)}{|\mathbf{G}|R_\alpha} e^{-i\mathbf{G}\cdot\mathbf{r}_\alpha} & |\mathbf{G}| \le G_{max} \\ 0 & |\mathbf{G}| > G_{max}. \end{cases} \tag{5.41}$$

Note that, since $\tilde{\Theta}(\mathbf{G})$ is to be used for calculating matrix elements, the appropriate G_{max} is twice the cutoff used for the basis functions.

Thus, the contributions to the overlap are

$$\frac{1}{\Omega}\int_{\Omega} \mathrm{d}^3\mathbf{r}\; e^{i(\mathbf{G}'-\mathbf{G})\cdot\mathbf{r}}\, \Theta(\mathbf{r}) = \tilde{\Theta}(\mathbf{G}-\mathbf{G}'). \tag{5.42}$$

A parallel procedure may be used to calculate the corresponding component of the Hamiltonian.

$$\frac{1}{\Omega}\int_{\Omega} \mathrm{d}^3\mathbf{r} e^{i(\mathbf{G}'-\mathbf{G})\cdot\mathbf{r}} V_{PW}(\mathbf{r})\Theta(\mathbf{r}) = \tilde{V}_{PW}(\mathbf{G}-\mathbf{G}'), \tag{5.43}$$

where $\tilde{V}_{PW}(\mathbf{G})$ is given in momentum space by the convolution,

$$\tilde{V}_{PW}(\mathbf{G}) = \sum_{\mathbf{G}'} V_{PW}(\mathbf{G}')\Theta(\mathbf{G}-\mathbf{G}'). \tag{5.44}$$

Since $\tilde{V}_{PW}(\mathbf{G})$ must be calculated up to the same cut-off, G_{max} as $\tilde{\Theta}$, the argument, $(\mathbf{G}-\mathbf{G}')$ of Θ in Eqn. 5.44 extends to a higher cut-off of $2G_{max}$. $\tilde{V}_{PW}(\mathbf{G})$ may be evaluated directly using the upper half of Eqn. 5.41, extended to $2G_{max}$ and Eqn. 5.44.

There is, however, a more efficient procedure, illustrated in Fig. 5.7. A step function $\bar{\Theta}$, defined identically to $\tilde{\Theta}$, but with a larger cut-off of $2G_{max}$ is constructed. FFTs are used to transform both $\bar{\Theta}$ and V_{PW} to real space (*n.b.* a mesh with twice the usual density is needed because of the larger cut-off). $\bar{\Theta}$ and V_{PW} are then multiplied on the discrete real space mesh points. The product is then back transformed to momentum space. This product, truncated to G_{max} is exactly the $\tilde{V}_{PW}$ to be used analogously to $\tilde{\Theta}$ in calculating the contribution from the interstitial potential to $H_{\mathbf{G},\mathbf{G}'}$.

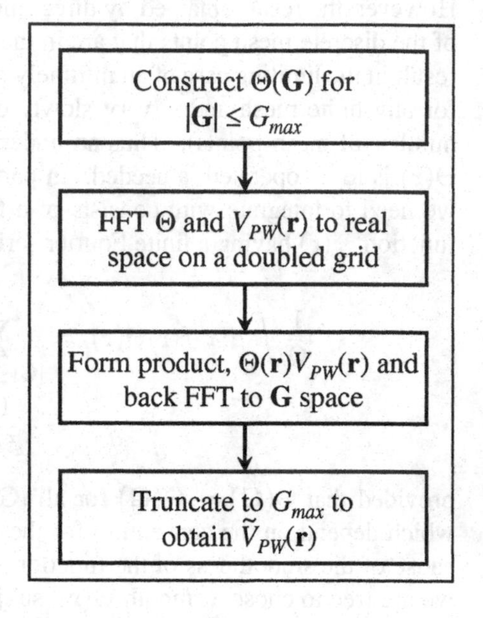

Figure 5.7. Computation of $\tilde{V}_{PW}$ using fast Fourier transforms.

The remaining interstitial term is the kinetic energy contribution to $H_{\mathbf{G},\mathbf{G}'}$. This may be obtained in terms of the step function, $\tilde{\Theta}$, using

$$\begin{aligned}\frac{1}{\Omega}\int_{\Omega} d^3\mathbf{r}\Theta(\mathbf{r})e^{-i(\mathbf{k}+\mathbf{G})\cdot\mathbf{r}}(-\nabla^2)e^{i(\mathbf{k}+\mathbf{G}i')\cdot\mathbf{r}} \\ = (\mathbf{k}+\mathbf{G}')^2\Theta(\mathbf{G}-\mathbf{G}') \\ = (\mathbf{k}+\mathbf{G}')^2\tilde{\Theta}(\mathbf{G}-\mathbf{G}'),\end{aligned} \tag{5.45}$$

where the last equality arises from the constraint $|\mathbf{G}|, |\mathbf{G}'| \leq G_{max}$ in $H_{\mathbf{G},\mathbf{G}'}$. In practice, it is normally more convenient to use an equivalent explicitely Hermetian form of the kinetic energy, specifically, replacing $(\mathbf{k}+\mathbf{G})^2$ by $(\mathbf{k}+\mathbf{G})\cdot(\mathbf{k}+\mathbf{G}')$.

The Spherical Terms

The spherical terms, $S_\alpha(\mathbf{G}, \mathbf{G}')$ and $H_\alpha(\mathbf{G}, \mathbf{G}')$ may be evaluated directly using the $a_{\ell m}(\mathbf{G})$ and $b_{\ell m}(\mathbf{G})$ coefficients along with the definitions of u_ℓ and $\dot{u}_\ell$, as given in the previous section. In this way, separate integration of the large kinetic and potential terms near the nucleus is avoided, along with the numerical instabilities that can result. Thus, using the orthonormality of the $Y_{\ell m}$ and the normalizations (Eqns. 5.22 and 5.23),

$$S_\alpha(\mathbf{G}, \mathbf{G}') = \sum_{\ell m} [a^*_{\ell m,\alpha}(\mathbf{G}) a_{\ell m,\alpha}(\mathbf{G}') + b^*_{\ell m,\alpha}(\mathbf{G}) b_{\ell m,\alpha}(\mathbf{G}') |\dot{u}_{\ell,\alpha}|^2], \quad (5.46)$$

where for brevity we use $\mathbf{G}$ as denoting $(\mathbf{k} + \mathbf{G})$. Similarly, using Eqns. 5.20 and 5.21,

$$\begin{aligned} H_\alpha(\mathbf{G}, \mathbf{G}') = \sum_{\ell m} &[a^*_{\ell m,\alpha}(\mathbf{G}) a_{\ell m,\alpha}(\mathbf{G}') \\ &+ b^*_{\ell m,\alpha}(\mathbf{G}) b_{\ell m,\alpha}(\mathbf{G}') |\dot{u}_{\ell,\alpha}|^2] E_{\ell,\alpha} \\ &+ \frac{1}{2}[a^*_{\ell m,\alpha}(\mathbf{G}) b_{\ell m,\alpha}(\mathbf{G}') + b^*_{\ell m,\alpha}(\mathbf{G}) a_{\ell m,\alpha}(\mathbf{G}')], \end{aligned} \quad (5.47)$$

where the final term was made explicitly Hermetian. These expressions may be evaluated efficiently by first synthesizing and storing the $a_{\ell m}$ and $b_{\ell m}$, and subsequently forming the outer products implicit in Eqns. 5.46 and 5.47 and adding the resulting contributions to $\mathbf{H}$ and $\mathbf{S}$.

The Non-Spherical Matrix Elements

The non-spherical components of $\mathbf{H}$ are integrals of wavefunctions with the non-spherical potential, which is given in a lattice harmonic expansion. Thus they are written in terms of radial integrals of the augmenting radial functions with the $V_{\nu\alpha}$; these are precomputed and then multiplied by the coefficients from the matching at the sphere boundary. In the non-relativistic case they are

$$I^{uu}_{\ell\ell',\nu\alpha} = \int_0^{R_\alpha} u_{\ell\alpha}(r) V_{\nu\alpha}(r) u_{\ell'\alpha}(r) r^2 \mathrm{d}r, \quad (5.48)$$

$$I^{u\dot{u}}_{\ell\ell',\nu\alpha} = \int_0^{R_\alpha} u_{\ell\alpha}(r) V_{\nu\alpha}(r) \dot{u}_{\ell'\alpha}(r) r^2 \mathrm{d}r, \quad (5.49)$$

and

$$I^{\dot{u}\dot{u}}_{\ell\ell',\nu\alpha} = \int_0^{R_\alpha} \dot{u}_{\ell\alpha}(r) V_{\nu\alpha} u(r) \dot{u}_{\ell'\alpha}(r) r^2 \mathrm{d}r, \tag{5.50}$$

where only the cases $\ell' \leq \ell$ need be computed in the symmetric terms (Eqns. 5.48 and 5.50). The scalar relativistic forms of the integrals are similar in form but involve both the large and small components of the radial functions. For example,

$$I^{uu}_{\ell\ell',\nu\alpha} = \int_0^{R_\alpha} [g_{\ell\alpha}(r) g_{\ell'\alpha}(r) + f_{\ell\alpha}(r) f_{\ell'\alpha}(r)] V_{\nu\alpha}(r) r^2 \mathrm{d}r. \tag{5.51}$$

In terms of these, the required matrix elements are

$$\begin{aligned} V_\alpha^{NS}(\mathbf{G},\mathbf{G}') = & \sum_{\ell m\, \ell' m'} \sum_{\nu} [a^*_{\ell m,\alpha}(\mathbf{G}) a_{\ell' m',\alpha}(\mathbf{G}') I^{uu}_{\ell\ell',\nu\alpha} \\ & + a^*_{\ell m,\alpha}(\mathbf{G}) b_{\ell' m',\alpha}(\mathbf{G}') I^{u\dot{u}}_{\ell\ell',\nu\alpha} + b^*_{\ell m,\alpha}(\mathbf{G}) a_{\ell' m',\alpha}(\mathbf{G}') I^{u\dot{u}}_{\ell'\ell,\nu\alpha} \\ & + b^*_{\ell m,\alpha}(\mathbf{G}) b_{\ell' m',\alpha}(\mathbf{G}') I^{\dot{u}\dot{u}}_{\ell\ell',\nu\alpha}] \int \mathrm{d}^2\omega Y^*_{\ell m}(\hat{\mathbf{r}}) Y_{\ell' m'}(\hat{\mathbf{r}}) \mathbf{K}_{\nu,\alpha}(\hat{\mathbf{r}}), \end{aligned} \tag{5.52}$$

where the $\mathbf{K}_{\nu,\alpha}$ are the lattice harmonics (Eqn. 5.5). The integral over solid angle, ω leads to a simplification in the sum over m in the definition of the lattice harmonics. This is because the Gaunt coefficients,

$$G_{\ell\ell'\ell'' mm'm''} = \int \mathrm{d}^2\omega Y^*_{\ell m}(\hat{\mathbf{r}}) Y_{\ell' m'}(\hat{\mathbf{r}}) Y_{\ell'' m''}(\hat{\mathbf{r}}), \tag{5.53}$$

are non-zero only if $m = m' + m''$. (Other conditions are that ℓ, ℓ' and ℓ'' must satisfy the triangle inequality, $|\ell' - \ell''| \leq \ell \leq \ell' + \ell''$, and that $\ell + \ell' + \ell''$ must be even.) Thus,

$$\int \mathrm{d}^2\omega Y^*_{\ell m}(\hat{\mathbf{r}}) Y_{\ell' m'}(\hat{\mathbf{r}}) \mathbf{K}_{\nu,\alpha}(\hat{\mathbf{r}}) = c_{\nu,\alpha,m-m'} G_{\ell\ell'\ell_\nu mm'(m-m')}. \tag{5.54}$$

A cursory inspection of Eqn. 5.52 suggests that the evaluation of the non-spherical contribution to the Hamiltonian can be very time consuming, and accordingly care is needed in the implementation of this step. An efficient approach is as follows (Fig. 5.8). The outer loops are over α, ℓ, m, ℓ' and m'. Inside these, the Gaunt coefficient for each lattice Harmonic is evaluated and, if any are non-zero, the contribution is added as the sum of two outer products (often very efficient, especially on pipelined and vector machines) of the form $a^*_{\ell m}(\mathbf{G}) A_{\ell m \ell' m'}(\mathbf{G}')$ and $b^*_{\ell m}(\mathbf{G}) B_{\ell m \ell' m'}(\mathbf{G}')$, where $A_{\ell m \ell' m'}$ and

$B_{\ell m \ell' m'}$ are defined by Eqn. 5.52 (note that the implied summation over ν should be done in the construction of the $A_{\ell m \ell' m'}$ and $B_{\ell m \ell' m'}$ and not in the outer product).

Even with careful coding, this step can be time consuming, often second only to the diagonalization of the Hamiltonian. Thus it is important to carefully chose the angular momentum cut-offs for the augmenting functions and the lattice harmonic expansion of the potential as small as possible, while retaining sufficient accuracy.

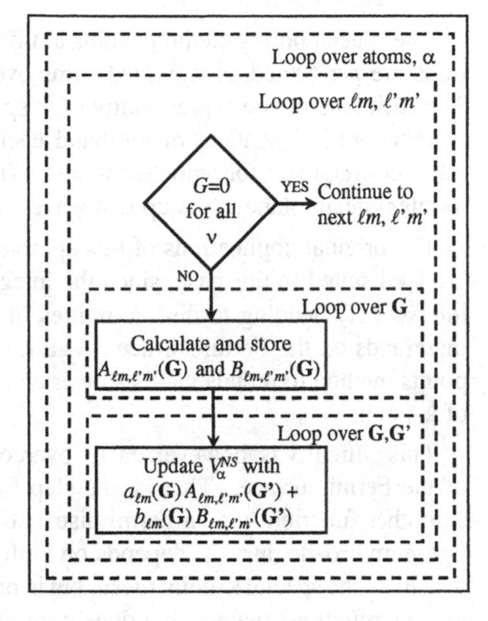

Figure 5.8. Evaluation of the non-spherical contribution to the Hamiltonian. The work inside the main loop is bypassed when the Gaunt coeffi cients (G) are zero.

5.6 Brillouin Zone Integration and the Fermi Energy

In order to determine the charge density and other quantities (total energy, forces, *etc.*), it is necessary to evaluate sums over the occupied states, which for crystals become integrals over the Brillouin zone, and, using symmetry, reduce to integrals over the irreducible wedge of the zone (IBZ). These integrals are necessarily calculated numerically using wavefunctions and eigenvalues at a finite number of **k**-points in the zone. The two commonly used approaches are the tetrahedron method [104, 105, 77, 50, 153] and the special points method [7, 33, 118, 119].

The Tetrahedron Method

In the tetrahedron approach, the zone is divided into tetrahedra and the band energies and wavefunctions are computed at **k**-points on the vertices (and additional locations for higher order methods). The band energies are interpolated (most often a linear interpolation is used, in which case the scheme is referred to as the linear tetrahedron method) between the vertices, and this interpolation is used to determine the Fermi energy, E_F. States below E_F are occupied and those above are unnoccupied. Each band at each **k**-point is then assigned a weight based on the volume of the tetrahedron for which it occurs below E_F, and the charge density is calculated by summing the individual contributions with these weights. The reader is referred to the references for further details.

The Special Points Method

The special points method adopts a different point of view in which integrations are performed as weighted sums over a grid of representative **k**-points. The locations of the representative (or special) points and the corresponding weights are independent of the band energies, and are chosen to yield optimum convergence for smooth functions (the points and weights are designed to integrate all planewaves up to a wavevector cut-off exactly).

The original applications of this approach were to insulators; special points are well suited to this case, since the integrands are smooth. In metals, bands intersect E_F leading to discontinuities in the occupation and therefore in the integrands on the Fermi surface. As a result direct, application of the special points method to metals yields slower convergence with respect to the number of **k**-points.

This difficulty is, however, easily overcome by using an artificial broadening of the Fermi surface. That is, the step function occupation is replaced by a smoother function, *e.g.* a Fermi distribution at some finite temperature. The optimum broadening, Δ, depends on both the band structure near E_F and the density of the special **k**-point mesh, but in practice a suitable value can be chosen using a rough estimate of the density of states at E_F, $N(E_F)$, the number of inequivalent **k**-points, n_k and the notion that to be effective the broadening should be such that it smooths on an energy scale given by the spacing of the eigenvalues, *i.e.* $\Delta \sim [n_k N(E_F)]^{-1}$. Of course, the Fermi surface of metals, and the discontinuity in occupation at E_F is physical, and so care needs to be taken in practical calculations to ensure that the broadening used to accelerate convergence does not affect physical quantities of interest.

The generation of a set of special points and weights for a general lattice [118, 119] proceeds as sketched in Fig. 5.9 (see below for two special cases). A grid is constructed in the full Brillouin zone using given divisions of the reciprocal lattice vectors; the grid is chosen so that it is offset from Γ by 1/2 division in each direction. Sets of symmetry related **k**-points are identified by applying the rotational parts of the group operations. This may be done, as illustrated, by rotating the points into the irreducible wedge, or by sequentially applying the space group operations to each **k**-point (this does not require knowledge of the IBZ). One representative **k**-point is chosen from each set of equivalent points, and a weight, $w(\mathbf{k})$, is assigned equal to the number of points in the set divided by the total number of points in the grid. These are the special points and associated weights.

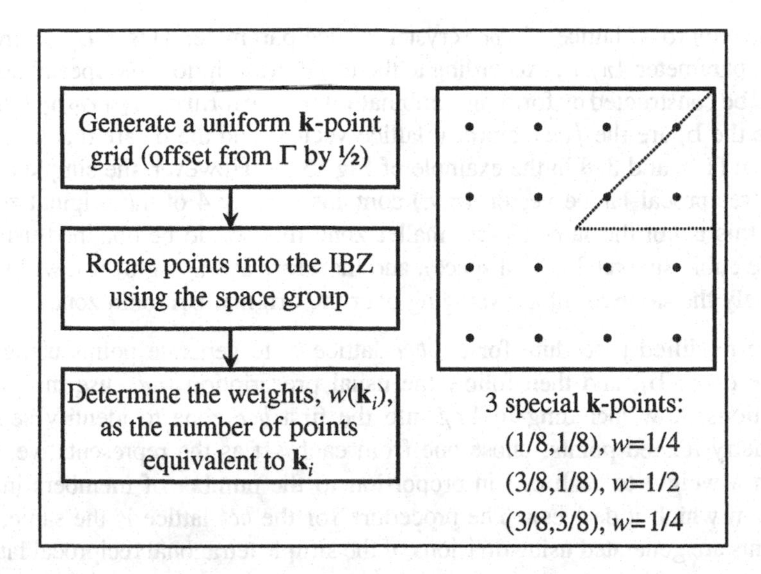

Figure 5.9. The generation of special **k**-points and weights. The right side gives an example for a 2-dimensional square lattice with divisions of $1/4$ of the zone.

Normally, using the smallest number of k-points for a given level of convergence is the main factor in the selection of a zone averaging scheme. The special points method with temperature broadening has a distinct advantage over the simplest (and most common) implementation of the linear tetrahedron method. This is because the set of special points generally includes fewer high symmetry points than the uniform, commensurate, Γ centered grids used with the tetrahedron method. Eigenvalues and wavefunctions at symmetry points contain less information regarding the dispersions than those at general points. This, in fact, was the motivation of the original formulation of the special points method by Baldereschi [7].

Special Treatment for Body Centered Lattices

Better convergence for body centered cubic (*bcc*) lattices may be obtained by modifying the procedure used to generate special points. A related modification has proved to be extremely valuable in studies of body centered tetragonal (*bct*) materials like some of the high critical temperature (T_C) superconducting cuprates. Chadi and Cohen [33] initially proposed use of a set of special points different that generated by the above scheme for certain lattices, based on the improved sampling of smooth functions obtained with the new set. The reason for this will be discussed in terms of the lattices after a description of the modification.

The reciprocal lattice of a *bcc* crystal (lattice parameter a) is fcc (reciprocal lattice parameter $4\pi/a$). According to the usual prescription, the special points would be constructed by forming combinations of the form $\alpha_1\mathbf{b}_1+\alpha_2\mathbf{b}_2+\alpha_3\mathbf{b}_3$ where the $\mathbf{b}_i$ are the fcc reciprocal lattice vectors and the α_i are the divisions of them ($1/8$ and $3/8$ in the example of Fig. 5.9). However, the simple cubic zone (reciprocal lattice vector $4\pi/a$) contains exactly 4 of the original zones (*n.b.* this is not the same as the smaller zone that would be obtained using a simple cubic supercell in real space), and therefore averaging over it will yield precisely the same result as averaging over the original Brillouin zone.

The modified procedure for the *bcc* lattice is to generate points using the simple cubic $\mathbf{b}_i$, and then follow the usual prescription (*i.e.* use the group operations, now including folding into the first *bcc* zone to identify sets of symmetry related points; chose one from each set as the representative, and assign a weight to each one in proportion to the number of members in the set from which it derives). The procedure for the *bct* lattice is the same, *i.e.* **k**-points are generated using divisions of the simple tetragonal reciprocal lattice vectors.

The reason that this modification is desirable for *bcc* lattices is that it generates fewer high symmetry **k**-points. In particular, no **k**-points of the form $(\alpha,\beta,0)$ are generated with the modification. The same advantage is present in the *bct* case, and for highly anisotropic systems such as the high T_c cuprates there is an additional benefit [36]. These materials consist of stacks of layers. The in-plane lattice parameters are much shorter than the out-of-plane lattice parameter and besides there a fewer efficient hopping channels between layers. Corresponding to this the band structures are almost two dimensional, with very little dispersion in the c-direction.

In this case, a fairly dense **k**-point mesh may be required along the in-plane directions, but not in the c-direction. However, in many of these materials, the layer stacking yields a *bct* lattice. Direct application of the standard unmodified procedure will yield an unnecessarily dense mesh along the c-direction for the following reason. Consider, the full symmetry ($I4/mmm$) *bct* lattice, and a set of mesh points $\{(a,b,c), (b,a,c), (-a,b,c), (a,-b,c), (-b,a,c), (b,-a,c), (-a,-b,c), (-b,-a,c)\}$ in *bct* reciprocal lattice coordinates. These **k**-points are not all symmetry related; if the group operations are used to rotate them into the IBZ they will have the same Cartesian k_x and k_y coordinates but different heights above the $k_z = 0$ plane, contrary to the goal of not wasting effort by sampling the out of plane direction. This happens because the $\mathbf{b}_1$ and $\mathbf{b}_2$ reciprocal lattice vectors have out of plane components. If, on the other hand, the mesh is constructed using the larger simple tetragonal zone with a coarse c-axis division, this coarseness will be preserved in the special points set (the in-plane reciprocal lattice vectors are now perpendicular to the tetragonal axis). A similar scheme can be applied to layered materials with orthorhombic symmetry, such as layered perovskite with octahedral tilts.

Special Treatment for Reduced Symmetry Lattices

There is at least one other case for which special treatment is desirable. This is for lattices that are distortions of a higher symmetry lattice, *e.g.* when calculating elastic constants or phonon frequencies from total energy variations (frozen phonon method). As an illustration, consider calculation of the total energy of a crystal with an fcc Bravais lattice.

The lattice vectors in this case are (0,1/2,1/2), (1/2,0,1/2) and (1/2,1/2,0) in units of a, and the reciprocal lattice vectors are (-1,1,1), (1,-1,1) and (1,1,-1) in units of $2\pi/a$ Applying the standard procedure, a small set of special **k**-points is (1/4,1/4,1/4) with weight 1/4 and (3/4,1/4,1/4) with weight 3/4, in the same units. Now consider this lattice with the symmetry lowered through a distortion along a cube diagonal (say [111]); this lowers the symmetry to trigonal. Applying the standard procedure to this lattice with the same divisions yields the set (1/4,1/4,1/4) with weight 1/4 and (-3/4,1/4,1/4) with weight 3/4. This set, however, does not have the symmetry of the cubic lattice, and this has an important consequence. If the energy is calculated as a function of the distortion, its extremum will be displaced from zero. Moreover, other quantities, such as forces and electric field gradients will fail to satisfy the cubic symmetry at zero distortion. In general, this problem occurs when the **k**-point grid before folding into the IBZ (first step of Fig. 5.9) has less symmetry than the space group, and the distortion destroys one or more of the lattice symmetries that are not present in the grid.

Of course, as the density of the **k**-point mesh is increased, both **k**-point sets (cubic and trigonal) will converge to the same result for zero distortion. Nonetheless, the use of these trigonal **k**-points is undesirable because of the cost imposed by the slow convergence to the correct symmetry for the undistorted lattice. Further, there is a simple solution, *i.e.* to use special **k**-points with the cubic symmetry. These may be generated in two ways, both amounting to enforcing the cubic symmetry on the unfolded **k**-points for the undistorted lattice.

The first method begins with the special **k**-points for the original higher symmetry (say cubic) lattice. These are spread over the full zone by application of the cubic group operations. (Equivalently, one could start with the special points of the lower symmetry, say trigonal, lattice, and apply the cubic operations.) Finally, the trigonal group operations are used as in the standard procedure to identify symmetry related points. One representative point is chosen from each set with a weight proportional to the sum of the weights of the **k**-points comprising the set. This procedure yields five special points for the example above: (1/4,1/4,1/4), (-1/4,1/4,1/4), (3/4,1/4,1/4), (3/4,-1/4,1/4) and

(-3/4,1/4,1/4). The reader may note that these **k**-points, although listed here in scaled Cartesian coordinates for clarity, should be stored in reciprocal lattice coordinates (coefficients of the $\mathbf{b}_i$) because then the coefficients and weights will then be invariant if the lattice vectors are strained (*e.g.* in the calculation of elastic constants).

The second procedure is parallel to the treatment for body centered lattices. A larger zone is formed using linear combinations of the reciprocal lattice vectors, such that the grid defined in terms of these new vectors has the full symmetry. In our example, this would be a simple cubic zone, containing two of the original zones. The special **k**-points are then constructed using this zone as in the discussion for body centered lattices, above.

Determination of the Fermi Energy

The Fermi energy, using special points and temperature broadening (see the references on the tetrahedron method for details when using that scheme), is determined by enforcing

$$\sum_{\mathbf{k},j} w(\mathbf{k}) F(\epsilon_{\mathbf{k},j}, E_F, T) = n_{tot}, \tag{5.55}$$

where the sum is over the special **k**-points and bands (j), n_{tot} is the total occupation number (the number of valence electrons for spin-polarized calculations, and half the number of electrons otherwise), the $\epsilon_{\mathbf{k},j}$ are the band energies, and F is a broadened occupation function, typically a Fermi function, with width Δ controlled by a temperature T,

$$F(\epsilon, E_F, T) = [\exp(\frac{\epsilon - E_F}{kT}) + 1]^{-1}. \tag{5.56}$$

Solution of Eqn. 5.55 proceeds via repeated evaluation of its left hand side, which is a monotonic function of E_F. The simplest (and quite effective approach) is to bracket E_F from above and below, and repeatedly narrow the brackets by evaluating the sum at the midpoint, and based on this evaluation moving either the upper or lower bracket to this position. This is the so-called bisection method. Alternate, efficient approaches are given for example in the book of Press *et al.* [150].

5.7 Computation of the Valence Charge Density

The valence charge density consists of two components. These are the interstitial charge, represented in stars, and the sphere charges, represented in lattice harmonics on radial grids. These are discussed in turn.

The Interstitial Density

The interstitial charge density is given by

$$\rho(\mathbf{r}) = \sum_s a_s \phi_s(\mathbf{r}) = \sum_{\mathbf{G}} c_{\mathbf{G}} e^{i\mathbf{G}\cdot\mathbf{r}}$$
$$= \sum_{\mathbf{k},j} W(\mathbf{k},j) \sum_{\mathbf{G},\mathbf{G}'} \phi^*_{\mathbf{G}',\mathbf{k},j} \phi_{\mathbf{G},\mathbf{k},j} e^{i(\mathbf{G}-\mathbf{G}')\cdot\mathbf{r}}, \qquad (5.57)$$

where $\mathbf{r}$ is restricted to the interstitial, the a_s are the star coefficients, the $c_{\mathbf{G}}$ are the corresponding planewave coefficients, the $\phi_{\mathbf{G},\mathbf{k},j}$ are the band eigenvector coefficients, j are the band indices, W is a weight that includes both the $\mathbf{k}$-point weight and an occupation factor (*e.g.* the Fermi factor), and the sum over $\mathbf{k}$ is over the full Brillouin zone. Note that, if the sum over $\mathbf{k}$ is restricted to the IBZ, the result of the final stanza of Eqn. 5.57 will not in general have the required full lattice symmetry.

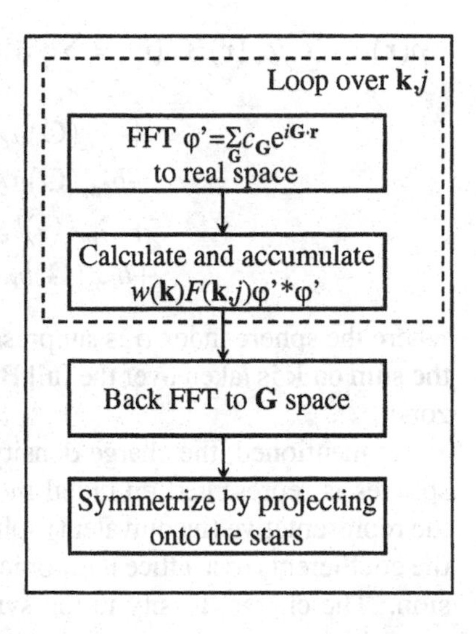

Figure 5.10. Computation of the interstitial charge density. The $\mathbf{k}$-point sum is over the IBZ and the symmetrization is done by projecting onto stars.

The procedure for calculating the a_s, which specify the interstitial charge density is as shown in Fig. 5.10. Since the eigenvectors are calculated only in the IBZ, symmetrization is required. This is done by projecting onto the stars (recall that these have the lattice symmetry). Moreover, since both planewaves and the stars are orthogonal functions, the projection is straightforward. The (unsymmetrized) $c_{\mathbf{G}}$ are determined using the last two stanzas of Eqn. 5.57 but with the sum over $\mathbf{k}$ restricted to the IBZ. The symmetrized a_s are formed as,

$$a_s = \sum_m \varphi^*_m c_{\mathbf{G}}, \qquad (5.58)$$

where the φ_m are the phases (Eqn. 5.1), and the sum is over star members, s.

Finally, the reader may note that the double sum on G and G' that appears in the last stanza of Eqn. 5.57 can be avoided by transforming the interstitial wavefunction (without the $\exp(i\mathbf{k}\cdot\mathbf{r})$ factors, which cancel) to real space via an FFT and forming the density by the multiplication $\phi^*\phi$ on this mesh. This greatly improves the efficiency of the calculation.

The Sphere Density

The approach for the charge density in the spheres is similar in spirit to that of the interstitial density. We begin with the general expression involving a sum over the whole Brillouin zone, but then reduce it to a sum over the IBZ, with symmetrization by projections onto appropriate functions. The charge density in a sphere is given by an expression related to that for the interstitial charge,

$$\begin{aligned}\rho(\mathbf{r}) = \sum_{\nu} \rho_{\nu}(r)\mathbf{K}_{\nu}(\hat{\mathbf{r}}) = \sum_{\mathbf{k},j} W(\mathbf{k},j) \sum_{\mathbf{G},\ell m} \sum_{\mathbf{G}',\ell' m'} \\ \{a^*_{\ell m}(\mathbf{G}) a_{\ell' m'}(\mathbf{G}') u_{\ell}(r) u_{\ell'}(r) \\ + b^*_{\ell m}(\mathbf{G}) a_{\ell' m'}(\mathbf{G}') \dot{u}_{\ell}(r) u_{\ell'}(r) \\ + a^*_{\ell m}(\mathbf{G}) b_{\ell' m'}(\mathbf{G}') u_{\ell}(r) \dot{u}_{\ell'}(r) \\ + b^*_{\ell m}(\mathbf{G}) b_{\ell' m'}(\mathbf{G}') \dot{u}_{\ell}(r) \dot{u}_{\ell'}(r)\} Y^*_{\ell m}(\hat{\mathbf{r}}) Y_{\ell' m'}(\hat{\mathbf{r}}),\end{aligned} \tag{5.59}$$

where the sphere index α is suppressed and the sum on $\mathbf{k}$ is taken over the full Brillouin zone.

As mentioned, the charge density in the spheres is represented on radial meshes in the representative (inequivalent) spheres by the coefficients of a lattice harmonic expansion. The charge density in the symmetry related atoms is the same as in the corresponding representative atom, aside from a rotation given by the space group operation that generates the equivalent atom from the representative atom. Since the local rotated coordinate frame is generally used in defining the lattice harmonics for equivalent atoms, the expansion of the charge density is identical in the equivalent and representative atoms. As in the case of the interstitial density, the sphere densities are constructed from the band eigenvectors in the IBZ. The required symmetrization is performed by projection onto the lattice harmonic representation. This is facilitated by the orthogonality of the lattice harmonics.

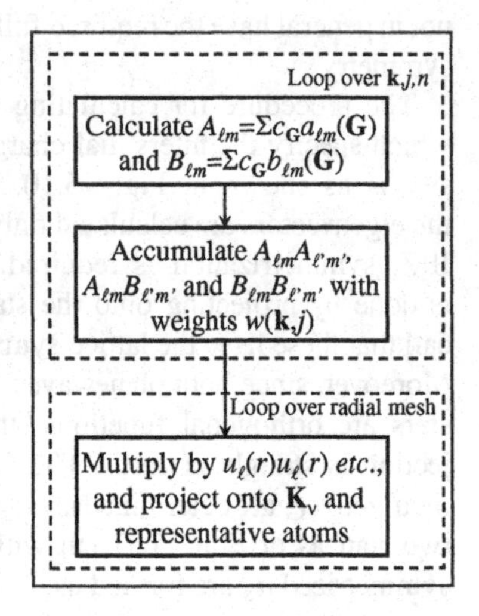

Figure 5.11. The sphere charges. The symmetrization projects the charge in each atom onto the representative atom, so the the loop on n is on all atoms.

The projection, P_ν of a term, $Y^*_{\ell m}(\hat{\mathbf{r}})Y_{\ell' m'}(\hat{\mathbf{r}})$ onto $\mathbf{K}_\nu(\hat{\mathbf{r}})$ is given in terms of the Gaunt coefficients (Eqn. 5.53) by

$$P_\nu = c^*_{\nu,m-m'} G_{\ell'\ell\ell_\nu m'm\, m-m'}, \tag{5.60}$$

where the $c_{\nu,m}$ are as in Eqn. 5.5 and the atom index is suppressed. For high symmetry lattices, there are many combinations of ℓm and $\ell' m'$ that cannot contribute because of the restrictions for them to have non-zero Gaunt coefficients.

The calculation proceeds as sketched in Fig. 5.11. Pairs of ℓm and $\ell' m'$ are pre-screened and the P_ν are determined for those values that can contribute. Then, for each band, the coefficients, $A_{\ell m}$ and $B_{\ell m}$, of u_ℓ and $\dot{u}_\ell$, respectively, are found for each atom, using the $a_{\ell m}$ and $b_{\ell m}$ and summing over $\mathbf{G}$; the weighted (by $W(\mathbf{k}, j)$) bilinear coefficients are accumulated for those pairs that can contribute. As mentioned, the symmetrization requires projecting both onto the lattice harmonics (done by using the P_ν) and onto the representative atoms. The latter is done most conveniently at this stage by calculating the bilinear coefficients for each member of a set of equivalent atoms, but then accumulating them together as if they were derived from a single atom. This introduces a factor of the number of equivalent atoms, N_{eq} in the set, which must be divided out at the end. After accumulating the bilinear coefficients for each band and $\mathbf{k}$-point, the lattice harmonic representation, $\rho_\nu(r)$ is constructed for each radial mesh point, using the values of $u_\ell(r)$ and $\dot{u}_\ell(r)$, the P_ν and Eqn. 5.59.

5.8 Core State Relaxation and Atomic Charge Densities

The LAPW method is an all-electron method, not in the sense that the core states are treated in the same manner as the valence states, but rather in the sense that (1) the valence states are orthogonal to the underlying core states (*i.e.* the true wavefunction is sought as opposed to a pseudo-wavefunction), and (2) the core states are calculated self-consistently in the crystal potential. In particular, the valence states are expanded in LAPW basis functions using the crystal potential, while the core states are treated using a numerical basis in an atomic approximation, *i.e.* for the core states, the wavefunctions are constrained to be spherical and all off-site overlaps are neglected. The core states are calculated using an atomic code with the potential replaced by spherical part of the crystal potential in the sphere. (Note that a full relativistic, *i.e.* Dirac, code is needed for this step, since for all but the lightest atoms, spin-orbit effects for core states are too strong to treat perturbatively.) Thus, both the core and valence states are calculated self-consistently, the core states fully relativistically in a spherical approximation, and the valence states using the full potential.

If the core states were completely confined inside the spheres, this would be the end of the story. Unfortunately, this is not exactly true in practice, and because of the strong long range nature of the Coulomb potential, artificial imposition of this may lead to significant errors (see Wei *et al.* [203] for numerical tests). Thus it is desirable to extend the above procedure to handle core tails that extend outside the LAPW spheres. This is done by approximating the potential outside by a smooth continuation of that inside (needed for the atomic program), calculating the core states in this potential, and accumulating the resulting core charge density. The form of extended potential outside the sphere is not particularly critical, provided that it is reasonable, *i.e.* agrees with the spherical component of the potential in the sphere at the boundary and has a short range decay to the Fermi energy outside it. One possibility is to use an exponential decay with a reasonable decay length (*e.g.* a bohr radius). Cohen [36] has suggested using this form, with the decay length fixed by requiring continuity of the potential and its derivative at the boundary, and in fact this approach works very well in practice.

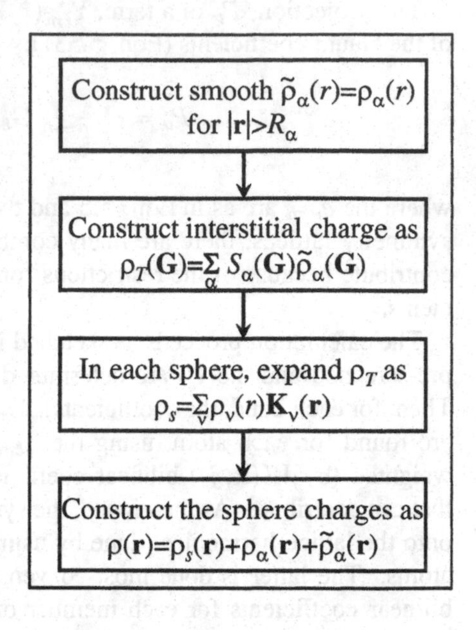

Figure 5.12. Conversion of extended core or atomic charge densities to the LAPW representation.

Once the core charge density outside a sphere has been found, it is necessary to convert its representation from values on a radial mesh to the LAPW representation of the charge density. An identical problem arises when initializing the charge density at the start of a calculation by overlapping the densities from atomic calculations, and so these are discussed together.

The fundamental problem is, given a spherically symmetric charge associated with each site, to convert it to the LAPW representation. This is done, as sketched in Fig. 5.12, using a technique very similar to the pseudo-charge technique used for the Coulomb potential. That is, the rapidly varying charge density is replaced by a smooth pseudo-charge that is correct in the interstitial. This is then used to construct the interstitial star representation as well as the expansion in the spheres due to neighboring atoms.

Specifically, each core (or atomic) density, $\rho_\alpha(\mathbf{r})$ is replaced by a pseudo-density, $\tilde{\rho}_\alpha\mathbf{r})$ that is equal to the original density for $r > R_\alpha$, but has a smooth continuation inside the sphere. This can be done using the same machinery that is used to construct the pseudo-charge for the solution of Poisson's equation (*n.b.* the multipoles don't matter in the present context). Next the Fourier transforms $\tilde{\rho}_\alpha\mathbf{G})$ are constructed. The interstitial component of the charge can now be synthesized as

$$\rho_T(\mathbf{G}) = \sum_\alpha S_\alpha(\mathbf{G})\tilde{\rho}_\alpha(\mathbf{G}), \tag{5.61}$$

where $S_\alpha(\mathbf{G}) = \exp(i\mathbf{G} \cdot \mathbf{r}_\alpha)$ is a structure factor, the sum is over atoms, α, and the $\mathbf{G}$ are reciprocal lattice vectors. The continuation of ρ_T into each sphere is then transformed into a lattice harmonic representation; this is done by re-expanding each planewave in Bessel functions (Eqn. 5.13), summing over $\mathbf{G}$ and projecting onto the lattice harmonics. The result of this operation contains the tails of the charge from neighboring sites as well as a spurious contribution, $\tilde{\rho}_\alpha(\mathbf{r}) - \rho_\alpha(\mathbf{r})$ due to the replacement with a pseudo-charge. This spurious contribution is subtracted from the density in each sphere to give the LAPW representation of the overlapped core or atomic charges as required.

5.9 Multiple Windows and Local Orbital Extensions

The LAPW basis set is constructed to be accurate for band energies near the linearization energies, E_ℓ (Andersen [2] presents a detailed discussion of the errors resulting from the linearization). In most materials, it is quite adequate to chose the E_ℓ near the center of the bands of interest. However, as mentioned, this is not always the case and there are important classes of materials for which there is no single choice of the E_ℓ that is adequate for all the bands that must be considered. A common example arises in $4f$ as well as early transition metal elements and compounds, where there are high lying and relatively extended core states (*e.g.* the $5p$ state in the $4f$ elements). Such states are also called semi-core states reflecting the fact that they are intermediate between band and core states. Often the spherical approximation inherent in the core treatment discussed above is inadequate, particularly in total energy calculations. Another example arises when bands extending over an unusually large region are required, as for example in comparing with experiments that probe high lying conduction bands. There are two common approaches for treating these situations: (1) the use of multiple energy windows and (2) relaxation of the linearization using local orbitals. These are discussed in turn.

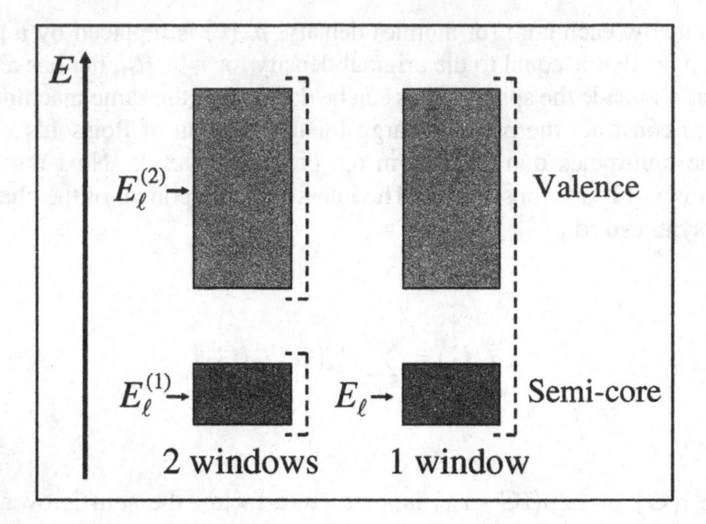

Figure 5.13. Example of windows with a semi-core state. The E_ℓ corresponding to the semi-core angular momentum is set low in the single window case.

Multiple Windows

Historically, the most common technique for dealing with these situations is to divide the energy spectrum into windows (sometimes called panels) and use a separate set of E_ℓ in each window. This procedure is illustrated for the treatment of a semi-core state in Fig. 5.13. However, we emphasize that the more recently developed local orbital technique is superior in almost all cases, and as such has largely replaced multiple window calculations.

In the two window treatment, a division is made into semi-core and valence energy regions. A different set of E_ℓ is chosen for each window to treat the states in it as accurately as possible. Separate calculations are then performed for the two windows and the relevant bands are used to construct semi-core and valence charge densities. This amounts to two independent LAPW calculations, with the exception that the potential is the same and the charge densities are combined.

Although this approach does solve many of the problems associated with the linearization and permits full relaxation of semi-core states, it is not fully satisfactory. First of all, there is a substantial overhead because separate calculations are being done for each window. When multiple windows are used for the conduction bands, the cost of a two window calculation is almost twice that of a one window calculation. For the more common application to semi-core states, the overhead is lower because these states are almost dispersionless and so only a very small number (typically only one) of **k**-points is needed for the lower window.

However, there are other problems with the treatment of semi-core states using multiple windows. In particular, the LAPW method relies on the fact that the radial functions, u_ℓ and $\dot{u}_\ell$ are orthogonal to any eigenstate that falls to zero at the sphere boundary, *i.e.* the underlying core states. However, semi-core states often satisfy this condition poorly even when the largest sphere radii consistent with the crystal structure are chosen. The result is that orthogonality of the valence wavefunctions to the semi-core state is not assured, unless the semi-core state is treated within the valence window (*i.e.* by setting the relevant energy parameter to the energy of the semi-core state). However, with this choice the calculation is equivalent to a one window calculation in that there is poor variational freedom for that ℓ character in the valence bands [115, 168].

To better illustrate the problem, we consider an example (Fig. 5.14) in which there is an extended semi-core p state and valence bands near the Fermi energy with some p character, as is typical of the early transition metals. If $E_{\ell=1}$ is set to the semi-core energy, ε_1, the semi-core states are well reproduced, and the valence bands are orthogonal to them (because of the diagonalization). However, the valence bands have poor variational freedom, and thus lie above their true positions.

As $E_{\ell=1}$ is raised the variational freedom of the semi-core states is rapidly degraded. This in itself is not a problem, since they could be treated in a separate window. However, as variational freedom is lost, ε_1 rises through the valence band region. It is often the case that energy parameters near the desired position in the valence bands yield degraded semi-core states (called ghost bands) overlapping the valence bands. Energy parameters for which ghost bands occur in the valence band region cannot be used because there is no good way of removing the ghost bands from the spectrum. Further, in this case the ghost and valence bands hybridize somewhat, degrading the quality of the valence bands. As $E_{\ell=1}$ is

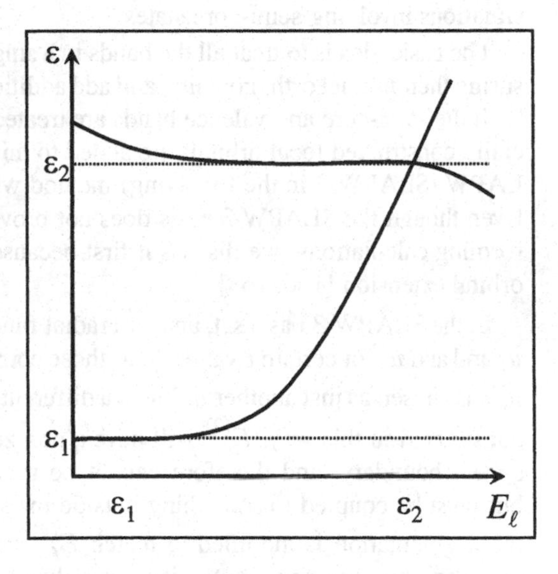

Figure 5.14. Variation of a semi-core and a valence band with the linearization energy, E_ℓ. The dotted lines at ε_1 and ε_2 denote the true locations of the bands.

further increased, the ghost bands leave the spectrum from above, but the valence bands lose their orthogonality to the semi-core states and the valence band energies begin to dip *below* their correct positions.

There are situations for which there is no adequate choice of energy parameters. These often arise in the early transition metals, especially when studying properties that depend on an accurate representation of the p-like character of the valence bands, *e.g.* some phonon frequencies [168] and electric field gradients [18, 171], as well as in $4f$ and $5f$ materials where the combination of unusually extended semi-core states and very soft lattices makes an accurate treatment of the valence p character particularly important [167, 52] .

Local Orbital Extensions

The local orbital extension of the LAPW method is a modification to the basis that circumvents the problems with multiple window calculations noted above. Although this approach does add some overhead to the calculation, in practice the overhead is considerably smaller than that added by using multiple windows. Besides avoiding the problems noted above for the relatively few cases where no choice of energy parameters in the standard LAPW method is adequate, it greatly simplifies the selection of energy parameters in other situations involving semi-core states.

The basic idea is to treat all the bands in a single energy window, thereby ensuring their mutual orthogonality, and add additional variational freedom so that both the semi-core and valence bands are treated accurately. In particular, specially constructed local orbitals are added to mimic the behavior of a modified LAPW (SLAPW-3 in the following) method with an enhanced augmentation. Even though the SLAPW-3 basis does not provide a practical scheme for performing calculations, we discuss it first because it leads naturally to the local orbital extension [166, 168].

In the SLAPW-3 basis set, an extra radial function, $u_\ell^{(2)}$ is added to the usual u_ℓ and and $\dot{u}_\ell$ for certain ℓ values (*e.g.* those corresponding to semi-core states); $u_\ell^{(2)}$ is chosen as just another u_ℓ but at a different energy parameter, $E_\ell^{(2)}$. When constructed in this way, $E_\ell^{(2)}$ will have a non-zero value and derivative on the sphere boundary, and therefore cannot be used directly as a basis function, but must be coupled to something outside the sphere. In the SLAPW-3 basis, the augmentation is modified to match $E_\ell^{(2)}$ to the planewaves on the sphere boundary by requiring continuity not only of the value and first derivatives of the basis functions, but the second derivatives as well. Related basis sets using various matching conditions have been proposed by Takeda and Kubler [191], Smrcka [180], Petru and Smrcka [144] and Shaughnessy *et al.* [160]. Particularly, SLAPW-3 is a special case of the basis proposed by Takeda and Kubler.

If E_ℓ is set to the energy of a semi-core state, a converged (with respect to the planewave cutoff) SLAPW-3 calculation has at least as good a basis as the LAPW method for that state. Further, if $E_\ell^{(2)}$ is chosen in the valence band region the SLAPW-3 basis will be quite good for the valence bands as well. Its variational freedom in the valence band region comes from $u_\ell^{(2)}$ and the linear combination of u_ℓ and $\dot{u}_\ell$ that is left over from the semi-core state. In another context, if $E_\ell^{(2)}$ is set very close to E_ℓ, the SLAPW-3 basis becomes the same as a quadratic APW method, using u_ℓ, $\dot{u}_\ell$ and the second energy derivative of u_ℓ, *i.e.* $\ddot{u}_\ell$, as may be seen by considering a finite difference expression for $\ddot{u}_\ell$. In the standard LAPW method, the error in a band energy, ε, varies as $(E_\ell - \varepsilon)^4$ while in this quadratic APW method the error is reduced to $(E_\ell - \varepsilon)^6$. Thus the SLAPW-3 method could in principle be used effectively to reduce linearization errors in the valence and conduction bands, thereby avoiding the need for multiple windows.

Unfortunately, the SLAPW-3 basis has a very serious drawback. Because of the requirement of continuous second derivatives, much higher planewave cutoffs are needed relative to the standard LAPW method to obtain a given level of convergence (recall that in Fourier space a first derivative brings down a factor of $|\mathbf{G}|$ while a second derivative brings down $|\mathbf{G}|^2$). Because of this, the SLAPW-3 method is not a practical scheme, except perhaps for the rare case where continuous second derivatives are required for some property calculation.

The widely applicable local orbital extension is constructed to retain the full variational freedom of the SLAPW-3 basis without requiring continuous second derivatives. As mentioned, this is done by starting with the standard LAPW basis: planewaves augmented by u_ℓ and $\dot{u}_\ell$ using continuity of the value and derivative on the sphere boundary. Local orbitals, consisting of linear combinations of u_ℓ, $\dot{u}_\ell$ and $u_\ell^{(2)}$, are added, with the particular linear combination determined by the conditions that the value and derivative of the local orbitals go to zero at the sphere boundary and choosing the coefficient of $u_\ell^{(2)}$ to some fixed number, say unity (a modification is used for lattices with inversion symmetry; see below). It can be shown (using the fact that the true wavefunctions do have continuous second derivatives) that the SLAPW-3 and LAPW plus local orbital basis sets give identical results in the large planewave cutoff limit. However, noting that in practice the number of local orbitals needed is much smaller than the number of planewaves (*e.g.* for s and p semi-core states, four local orbitals per atom would be used), the convergence of the local orbital extension is practically the same as that of the standard LAPW method. This is a substantial improvement over the SLAPW-3 basis.

One further modification is generally used for lattices that have inversion symmetry. In this case, the origin is normally chosen at an inversion center so that the secular equations become real. In order to exploit this with local orbitals, linear combinations that transform like planewaves must be used. These can be constructed simply by attaching the local orbitals to fictitious planewaves. That is, a set of planewaves is chosen and the corresponding structure factors, $F_{\ell m,\alpha}$ are constructed for each atom, α, as described above for the construction of the LAPW basis set (Eqns. 5.33 and 5.34). These are used as coefficients in the linear combinations. This requires use of N planewaves, where N is the total number of local orbitals in the basis, subject to the constraint that they yield linearly independent basis functions. Although not all sets of planewaves satisfy this constraint, finding such a set is straightforward, and can be done, *e.g.* by selecting planewaves one at a time and testing for linear independence of the corresponding combination of local orbitals with the previously selected members of the set.

5.10 The APW+LO Basis Set

It is clear that an energy independent APW basis set alone does not provide enough flexibility for finding solutions in a wider region around the fixed energy parameter. It was this understanding that paved the way for the LAPW method. However, from the experience with the above mentioned SLAPW-3 scheme versus local orbital extensions of this basis set, we might expect that this lack of variational freedom in the APW basis set can be compensated by using complementary local orbitals. In the spirit of the traditional LAPW basis set, these local orbitals should include the $\dot{u}_\ell(r)$ functions. These will then introduce the flexibility to replace the energy dependent basis set needed in the standard APW approach. Additionally, we would expect that this APW+LO basis set would gain in convergence in terms of number of basis functions, due to the less restricted augmentation compared to the traditional LAPW basis set. Only the value of the basis function is constant at the MT sphere instead of both value and slope.

Hence, this new basis set consists first of an energy independent APW part, as in Eqn. 4.1, but where u_ℓ is evaluated for a fixed energy $E = E_\ell$. In addition we have the local orbital extension, which now takes the form,

$$\mathcal{X}_L^{lo}(\mathbf{r},\mathbf{k}) = \begin{cases} 0 & \mathbf{r} \in \mathrm{I} \\ R_L^{lo}(r) Y_L(\hat{\mathbf{r}}) & \mathbf{r} \in \mathrm{MT}. \end{cases} \tag{5.62}$$

Here, $R_L^{lo}(r) = a_L^{lo} u_\ell(r, E_\ell) + b_L^{lo} \dot{u}_\ell(r, E_\ell)$. a_L^{lo} and b_L^{lo} are connected from the fact that $\mathcal{X}_L^{lo}$ should vanish at the MT-boundary.

The new basis-functions, here after referred to as APW+LO, will differ in some important aspects from the LAPW basis functions inside the MT-region. Most important is that APW+LO set uses $u_\ell(r, E_\ell)$ both in its original APW form and in a (less restricted) linear combination with $\dot{u}_\ell(r, E_\ell)$. Secondly, the linear combinations with $\dot{u}_\ell(r, E_\ell)$ need enter only for physically important ℓ-quantum numbers, *i.e.* for $\ell \leq 3$. Hence, the setup of matrix elements is faster using APW+LO, since the majority of the basis functions include only u_ℓ, and no $\dot{u}_\ell$. A third important difference between APW+LO and LAPW is that the APW+LO basis functions have no restriction on their derivatives at the MT-boundaries. The kinetic energy of the secular matrix, $T_{\mathbf{GG'}}$ must therefore be treated with special caution [176, 111, 174].

Once the standard local orbitals scheme is implemented into a LAPW code, the changes for adding the APW+LO basis are straightforward. All the machinery within the LAPW code, such as the full potential implementation, work equally well for APW+LO, with the exception that the kinetic energy needs to be treated differently.

The kinetic energy operator

According to Green's theorem, the two expressions for the kinetic energy operator $\int_V \mathcal{X}^*_{\mathbf{G}}(-\nabla^2)\mathcal{X}_{\mathbf{G'}}\,\mathrm{d}V$ and $\int_V (\nabla\mathcal{X}_{\mathbf{G}})^\dagger \cdot \nabla\mathcal{X}_{\mathbf{G'}}\,\mathrm{d}V$ are no longer equivalent when the functions $\mathcal{X}_{\mathbf{G}}$ have discontinuous derivatives. Instead they differ by an integral over the surface of discontinuity. Our surface of discontinuity is the sphere boundary, with $\mathrm{d}\mathbf{S}$ pointing outwards from the spheres. One clearly sees that the surface term vanishes if the functions are smooth.

$$\oint_{S_{\mathrm{MT}}} \mathcal{X}^*_{\mathbf{G}}\left(\frac{\partial\mathcal{X}^{\mathrm{MT}}_{\mathbf{G'}}}{\partial r} - \frac{\partial\mathcal{X}^{\mathrm{I}}_{\mathbf{G'}}}{\partial r}\right)\mathrm{d}\mathbf{S}. \tag{5.63}$$

While the Laplacian operator is commonly used to describe the kinetic energy operator, the formulation using nabla operators is symmetric with respect to the two functions $\mathcal{X}_{\mathbf{G}}$ and $\mathcal{X}_{\mathbf{G'}}$. Slater [176] argues that the second formulation is more fundamental as it enters already in the variational principle leading to the Schrodinger equation. Taking this standpoint, the extra term coming from Green's theorem must be added to the kinetic energy operator whenever the first formulation is used,

$$\begin{aligned} T_{\mathbf{GG'}} &= \int_{I+MT} \mathcal{X}^*_{\mathbf{G}}(-\nabla^2)\mathcal{X}_{\mathbf{G'}}\,\mathrm{d}V \\ &\quad + \oint_{S_{\mathrm{MT}}} \mathcal{X}^*_{\mathbf{G}}\left(\frac{\partial\mathcal{X}^{\mathrm{MT}}_{\mathbf{G'}}}{\partial r} - \frac{\partial\mathcal{X}^{\mathrm{I}}_{\mathbf{G'}}}{\partial r}\right)\mathrm{d}\mathbf{S}. \end{aligned} \tag{5.64}$$

Or, as the nabla operator is more convenient in the interstitial region,

$$\begin{aligned} T_{\mathbf{GG'}} &= \int_I (\nabla\mathcal{X}_{\mathbf{G}})^\dagger \cdot \nabla\mathcal{X}_{\mathbf{G'}}\,\mathrm{d}V \\ &\quad + \int_{\mathrm{MT}} \mathcal{X}^*_{\mathbf{G}}(-\nabla^2)\mathcal{X}_{\mathbf{G'}}\,\mathrm{d}V \\ &\quad + \oint_{S_{\mathrm{MT}}} \mathcal{X}^*_{\mathbf{G}}\frac{\partial\mathcal{X}^{\mathrm{MT}}_{\mathbf{G'}}}{\partial r}\,\mathrm{d}\mathbf{S}. \end{aligned} \tag{5.65}$$

A different approach, starting from the Laplacian operator is adapted by Schlosser and Marcus [159]. They then have to modify the variational expression, taking into account the discontinuity in slope at the MT-sphere, but end up with the same expression, Eqn. (5.64) for the kinetic energy operator.

5.11 Charge Density Mixing for Self-Consistency

The time required to perform a self-consistent calculation using the LAPW method is proportional to the number of iterations needed to reach self- consistency (see Fig. 2.1). Thus it is quite important to chose an efficient mixing of input and output charge densities. Straight mixing (Eqn. 2.18) is conceptually the simplest scheme, is easy to implement and does converge for small enough mixing parameters. Unfortunately, because all components of the density are treated equally in this method, the radius of convergence is fixed by the least well behaved mode. A common manifestation of this occurs in calculations for large unit cells (*e.g.* surface calculations using a supercell geometry). In this case, charge can oscillate between two parts of the cell (charge sloshing) from one iteration to the next. This long range (small $|\mathbf{G}|$) oscillation is strongly coupled to the electronic structure via the Coulomb potential, and very small mixing parameters are often needed to damp it. Another common example arises in magnetic systems, where there may be two or more solutions of the KS equations with similar energies (*e.g.* different spin configurations), and the charge and spin densities may bounce among them (see, *e.g.*, Ref. [39]). This problem becomes even worst in non-collinear magnetic calculations. On the other hand, there are other modes (*e.g.* short range charge oscillations in an atom), for which quite large mixing parameters would yield convergence. The typically large mismatch between the slowest converging mode and the fastest makes the straight mixing scheme inefficient. Accordingly, straight mixing is not generally used in electronic structure calculations. Instead modified Newton-Raphson methods, like Broyden's method [28] are generally used (*n.b.* in Car-Parrinello-like methods for insulators it is possible to avoid the need for mixing entirely by direct minimization of the Kohn-Sham Hamiltonian with respect to the orbital coefficients - see, *e.g.* Ref. [139]).

The basic idea behind Broyden's and related methods [195, 79] is to use information about the fluctuations occurring in previous iterations along with that from the current iteration to construct an input for the next. This means that if a mode is oscillating in previous iterations it can be damped and conversely if it is overdamped it can be mixed more strongly. This is accomplished by incorporating information from the current and previous iterations into an approximate inverse Jacobian, $G^{(m)}$, where m is the number of iterations; this is then used in updating the charge density.

$$\rho_{in}^{(m+1)} = \rho_{in}^{(m)} - G^{(m)} F^{(m)}, \tag{5.66}$$

with

$$F^{(m)} = \rho_{out}^{(m)} - \rho_{in}^{(m)}, \tag{5.67}$$

where $\rho^{(m)}$ is a vector of coefficients that define the charge density, and the subscripts in and out refer to input and output charge densities, respectively.

The dimension of these vectors in the LAPW method is quite large (see below) and this precludes storage of the matrices, $G^{(m)}$. For the first iteration, a guess, $G^{(1)}$ is used (say straight mixing). This is then refined as the iterations proceed. Thus the total information content in $G^{(m)}$ is at most that in $G^{(1)}$ plus that contained in $(m-1)$ input and output charge densities, *i.e.* a manageable amount. Therefore, the information needed for the updating is stored as an alternative to storing the $G^{(m)}$. In Broyden's method, the updating procedure is constructed to satisfy

$$G^{(m)}(F^{(m)} - F^{(m-1)}) = \rho_{in}^{(m)} - \rho_{in}^{(m-1)}. \tag{5.68}$$

This is the generalization to many dimensions of linearly extrapolating the change in a function, F, known at two values of its argument, ρ, to its zero. This yields a unique $G^{(m)}$ in one dimension; in higher dimensions, additional constraints are needed to fix $G^{(m)}$. Broyden's method provides these by requiring the change in the norm of the (estimated) Jacobian, (G^{-1}) to be minimal. This results in the updating scheme for G,

$$G^{(m)} = G^{(m-1)} + \frac{[\rho_{in}^{(m)} - \rho_{in}^{(m-1)} - G^{(m-1)}(F^{(m)} - F^{(m-1)})](F^{(m)} - F^{(m-1)})^T}{(F^{(m)} - F^{(m-1)})^T(F^{(m)} - F^{(m-1)})}, \tag{5.69}$$

where M^T denotes the matrix transpose of M. Direct implementation of the scheme embodied in Eqn. 5.69, within the LAPW method, is, however, impractical because it would involve the storage of $G^{(m)}$ in matrix form. However, Srivastava [186] rewrote the updating scheme (Eqn. 5.66) as

$$\rho_{in}^{(m+1)} = \rho_{in}^{(m)} - G^{(1)}F^{(m)} - \sum_{j=2}^{m} U^{(j)}V^{T(j)}F^{(m)}, \tag{5.70}$$

with

$$U^{(i)} = -G^{(1)}(F^{(i)} - F^{(i-1)}) + \rho_{in}^{(i)} - \rho_{in}^{(i-1)} - \sum_{j=2}^{i-1} V^{T(j)}(F^{(i)} - F^{(i-1)})U^{(j)}, \tag{5.71}$$

and

$$V^{T(i)} = \frac{(F^{(i)} - F^{(i-1)})^T}{(F^{(i)} - F^{(i-1)})^T(F^{(i)} - F^{(i-1)})}. \tag{5.72}$$

Broyden's method, when implemented in this way, requires the storage of just two vectors for each iteration, *i.e.* exactly the amount of information contained in the input and output charge densities. Since these vectors are accessed only once per iteration in a linear fashion, they may be stored conveniently. Moreover, in this form, all the operations have been expressed as vector-vector products, instead of the potentially costly matrix-vector operations that would be implied by a straight forward reading of Eqn. 5.69.

In the LAPW method, the $\rho^{(m)}$ are vectors of the real and imaginary components of the star coefficients concatenated with the values on the radial mesh of the $\rho_{\nu,\alpha}(r)$ for each lattice harmonic and atom. Broyden's method, with this representation, is quite effective in accelerating the convergence of self-consistent LAPW calculations [164]. Nonetheless, improvements are possible. Broyden's method weights information generated at the beginning of a self-consistent equally with that from the most recent iteration, even in the final stages of the calculation. In addition, information in $G^{(m)}$ from previous iterations is overwritten arbitrarily during standard Broyden's updates, and as a result, although the charge density converges, $G^{(m)}$ may not converge to the true inverse Jacobian as m grows large. Vanderbilt and Louie [195] proposed a modification of Broyden's method that weights the information from earlier steps differently using least squares to minimize the error in the inverse Jacobian. With a suitable choice of weights, this method both converges $G^{(m)}$ to the true inverse Jacobian, and yielded faster convergence in their LCAO calculations. Johnson [79], constructed a related approach that does not require large dimensioned matrix-vector operations and yielded good convergence in the context of KKR-CPA and LCAO calculations. However, although promising, tests of this approach have not been reported in the context of LAPW calculations, and accordingly, the interested reader is referred to Johnson's paper for details. Essentially, equations similar in form to those of Srivastava are obtained, but with coefficients that come from the inversion of $m \times m$ (m is the number of iterations) matrices (used to satisfy the constraints).

5.12 Fixed Spin Moment Calculations

Transition metals and their compounds display a particularly rich set of properties, many of which are related to magnetism and/or proximity to magnetic transitions. However, despite the fact that it is possible to investigate magnetic materials using standard self-consistent LSDA electronic structure approaches, much of the richness present in the magnetic phase diagrams of even simple transition metal elements and binary compounds was not appreciated until a series of seminal papers by Moruzzi and his co-workers [121, 122, 123, 124,

125, 126, 127, 128, 129, 130, 131]. These papers used the fixed spin moment (FSM) procedure [205, 158], which greatly simplifies the study of magnetic materials, particularly those with ferromagnetic phases. This section describes the procedure and its implementation.

The idea underlying the FSM approach is to do total energy calculations with the total moment constrained to fixed values. In this way the total energy as a function of the spin moment, M, can be mapped out. As Moruzzi and co-workers have shown, it is not at all unusual to have multiple minima in the $E(M)$ curve. These metastable states can be readily identified and studied using the FSM approach.

When combined with the conservation of total charge, the FSM constraint amounts to separately fixing the total up and down spin charges (the sum is the total charge and the difference is the spin moment). This is straightforward in practice (Fig. 5.15). A FSM calculation proceeds exactly as a standard LAPW calculation, except that separate Fermi energies are determined for the up- and down-spin channels, and these are then used in the separate construction of the up- and down-spin charge densities. The total energy calculation is exactly as in a standard calculation but with the eigenvalue sum including the eigenvalues occupied in the construction of the spin densities, *i.e.* up to the up- (down-) spin Fermi energy for spin-up (down) eigenvalues. This procedure is equivalent to performing calculations using constant stabilizing fields, B, and adjusting B to produce the desired value of the spin magnetization, M. Viewed in this way, B is just half the difference between the down- and up-spin Fermi energies.

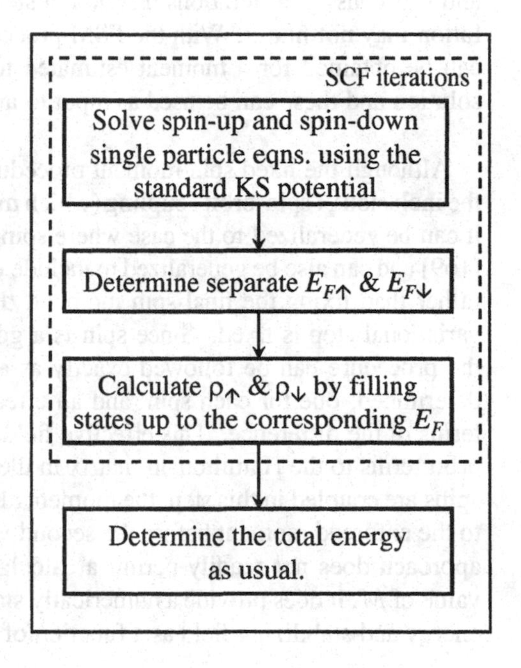

Figure 5.15. The fixed spin moment procedure.

Use of the FSM procedure offers several advantages in studying ferromagnetic materials. As mentioned, it allows the mapping out of $E(M)$, and thereby facilitates the identification of metastable ferromagnetic phases. Analysis of the FSM, $E(M)$ curves for materials near ferromagnetic quantum critical points has been used to good advantage in understanding the beyond LSDA quantum critical fluctuations [1, 101, 116]. Besides this, self-consistent iterations with the FSM method are much more stable than in standard spin-polarized calculations. In situations (*e.g.* Pd) where $E(M)$ is quite flat, charge can slosh between the two spin channels with very little energy cost in a standard calculation. Although convergence accelerators, like Broyden's method, do help in such situations, it can still be quite difficult to fully converge spin-polarized calculations, while FSM iterations converge as rapidly as non-spin-polarized calculations. Further, the FSM procedure is quite useful in obtaining good starting spin densities to be used in standard spin-polarized calculations. In many transition metal ferromagnets, the non-spin polarized solution is metastable, and self-consistent iterations that do not start close enough to the magnetic solution may not find it. With the FSM procedure, self-consistent spin densities can be obtained for a moment estimated to be close to that of the magnetic solution and these can be used as input to an unconstrained calculation.

Although the fixed spin moment procedure was formulated for use without the inclusion of spin-orbit coupling (which mixes spin-up and spin-down states), it can be generalized to the case where spin-orbit is included as a perturbation [169] and can also be generalized to include orbital moments [101]. In this case, rather than fixing the final spin moment, the spin moment before the second variational step is fixed. Since spin is a good quantum number at this point, the procedure can be followed exactly as above, *i.e.* two Fermi energies are determined, one for each spin, and an effective magnetic field, B, defined in terms of the difference. This effective field is then added along with the spin-orbit terms to the Hamiltonian matrix in the second variational step. Since the spins are coupled in this step, the moment although close to is not exactly equal to the enforced moment before the second variation. Although the generalized approach does not readily permit a calculation exactly at any predetermined value of M, it does provide a numerically stable prescription for calculating the energy and stabilizing field as a function of the magnetization.

5.13 The Total Energy

The total energy in the LAPW method is given by Eqn. 2.16, rewritten to cancel the divergent contributions from the ion-ion (E_{ii}) and Hartree terms. This is done by replacing V_H by the full Coulomb potential V_C in the last (double counting) term. The remainder of the ion-ion interaction is then a Madelung term with the divergence canceled. Also, since the cores are treated separately in the LAPW method, their contribution to the eigenvalue sum is separated. The resulting energy expression is,

$$E = \sum_{\alpha,c} \epsilon_{\alpha,c} + \sum_{occ} W(\mathbf{k}, j)\epsilon_{\mathbf{k},j} - \frac{1}{2}\sum_{\alpha} Z_\alpha R(\alpha)$$
$$- \int d^3\mathbf{r}\, \rho(\mathbf{r})[V_{xc}(\mathbf{r}) - \varepsilon_{xc}(\mathbf{r}) + \frac{1}{2}V_C(\mathbf{r})], \quad (5.73)$$

where the integral is over the unit cell, the sum over c is over the core states, ρ is the total charge density (valence and core), ε_{xc} is the exchange-correlation energy density per atom (Eqn. 2.5), Z_α is the nuclear charge on atom α and $R(\alpha)$ in the Madelung term is the Coulomb potential at the nucleus less the Z_α/r self-contribution. This last term is determined by taking the $\ell = 0$ Coulomb potential on the sphere boundary (*i.e.* $\rho_{\nu=0}$ on the last mesh point) and integrating Poisson's equation inside the sphere using Eqn. 5.19, but without including the nuclear charge in ρ_0 (note that it is included in the usual potential generation step). The same energy expression may be used in spin-polarized calculations with ρ now being the sum of the two spin densities, ε_{xc} now a function of the two local spin densities, and V_{xc} replaced by

$$V_{xc} \Rightarrow \frac{\rho_\uparrow V_{xc,\uparrow} + \rho_\downarrow V_{xc,\downarrow}}{\rho}, \quad (5.74)$$

where $\uparrow$ and $\downarrow$ denote spin-up and spin-down, respectively.

As written, the total energy expression is not variational. That is, if input densities and potentials are used, as is most convenient, the calculated values of the energy, E, during self-consistent iterations may fluctuate both above and below the final value. In planewave codes, it is conventional to use a variational expression for E because then the energy for any iteration forms an upper bound to the final energy. However, in LAPW codes this is not the norm because it is simpler to implement Eqn. 5.73 using only input quantities. Nonetheless, a variational expression can be obtained by replacing the charge density, ρ in the final (double counting) term by the output charge density, keeping V_{xc} and V_C from the input potential calculation, but calculating ε_{xc} using the output density.

5.14 Atomic Forces

Two basic applications of density functional methods are the determination of the ground state structures and calculation of vibrational modes of materials. In principle, both of these problems can be solved by mapping out the energy surface (the energy as a function of atomic positions) using total energy calculations, and this has been done with considerable success. In fact, this approach has been applied not only to simple materials, but also to relatively complex problems such as surface reconstructions and even phonon calculations for the high-Tc cuprate superconductors [37]. Nonetheless, this method becomes very inefficient as the number of atoms, N, in the unit cell grows. To illustrate this, consider a unit cell containing N atoms. In the simplest case, the energy surface is quadratic, and is described by coefficients that specify the minimum energy, relative equilibrium positions of the atoms, the equilibrium lattice parameters, and the Γ point dynamical matrix. Even in this simple case at least this number of total energy calculations would be needed to fully specify the energy surface for Γ point distortions. The resulting $O(N^2)$ scaling in the number of calculations needed (which gets even worst for more realistic anharmonic energy surfaces) combined with the $O(N^3)$ cost of a single total energy calculation, greatly restricts the applicability of approaches based only on total energy calculations. On the other hand, if forces on the nuclei as well as the energy are calculated, instead of one, $3N - 2$ independent pieces of information are obtained from each electronic structure calculation, in the form of an energy and $3N$ force components, which, however, must sum to zero in each direction. This yields a substantial reduction in the number of calculations that need to be performed if forces are available, and this advantage becomes greater as the complexity of the material increases. Accordingly, it is standard in many electronic structure methods to calculate and use atomic forces to relax the coordinates of atoms and calculate vibrational properties. Clearly, following the present trend towards materials by design and other applications of density functional calculations, where complex materials for which little if any experimental structural information is available, an efficient force calculation is crucial in any widely used methodology. However, despite this strong motivation, force calculations using the LAPW method appeared quite some time after the initial formulation of the method [181, 182, 211, 53]. This was a result of the complications that arise due to the use of a position dependent basis with discontinuous second derivatives in the LAPW method. This section discusses the calculation of forces in the LAPW method.

The modifications for the LAPW+LO basis are trivial, and involve just the addition of $c_{\ell m}u^{(2)}$ terms like the $a_{\ell m}u$ and $b_{\ell m}\dot{u}$ terms in the expressions. The APW+LO method involves higher order discontinuities on the sphere boundaries, but can be implemented following the same approach.

In quantum mechanical calculations with position independent basis sets (*e.g.* planewaves), the force on a nucleus is given by the Hellmann-Feynman theorem, and is exactly the electrostatic force acting on it in the self-consistent charge density [71, 46]. Additional terms, referred to as incomplete basis set (IBS) corrections, occur when incomplete basis sets that depend on the atomic positions (e.g. atomic orbitals) are used [151].

Bendt and Zunger [16] gave a general prescription for calculating forces in any basis set. The forces are based on the total derivative of the energy expression (including implicit dependencies) with respect to the atomic positions. They used this prescription to derive a compact expression for the IBS correction in the usual case where the only dependence of the Hamiltonian on the atomic positions is through the potential energy. However, the LAPW basis functions have discontinuities in their second derivatives; as conventionally implemented, this results in position dependent discontinuities in the kinetic energy. Starting with the total energy expression, Soler and Williams [181, 182] first derived and used formulae for the LAPW forces. Yu *et al.* [211] independently derived and used a different expression that isolated the terms from the discontinuous second derivatives. These two force expressions were subsequently shown to be equivalent [183]. Here the expression of Yu *et al.* is followed.

The force on an atom, α, is determined by the change in energy (Eqn. 5.73) when the atom is displaced by $\delta\mathbf{r}_\alpha$,

$$\delta E = \sum_{\mathbf{k},j} W(\mathbf{k},j)\delta\epsilon_{\mathbf{k},j} + \sum_{\alpha,c}\delta\epsilon_{\alpha c} - \int \mathrm{d}^3\mathbf{r}\rho(\mathbf{r})\delta V_{KS}(\mathbf{r}) - \mathbf{F}^\alpha_{HF}\cdot\delta\mathbf{r}_\alpha, \quad (5.75)$$

where

$$\mathbf{F}^\alpha_{HF} = Z_\alpha \frac{\mathrm{d}}{\mathrm{d}\mathbf{r}_\alpha}[-\sum_\beta{}' \sum_\tau \frac{Z_\beta}{|\mathbf{r}_\alpha - \mathbf{r}_\beta + \tau|} + \int \mathrm{d}^3\mathbf{r}\frac{\rho(\mathbf{r})}{|\mathbf{r}_\alpha - \mathbf{r}|}], \quad (5.76)$$

is the bare Hellmann-Feynmann force, V_{KS} is the Kohn-Sham potential, τ are direct lattice vectors, α and β are atom indices, and the integrals are over all space. $\mathbf{F}^\alpha_{HF}$ depends on the nuclear charge Z_α and the $\ell = 1$ part of the Coulomb potential near the nucleus.

$$\mathbf{F}^\alpha_{HF} = \sum_{\nu|\ell_\nu=1} \lim_{r\to 0} [\frac{V^C_\nu(r)}{r}] \sum_m \nabla[rY_{1m}(\hat{\mathbf{r}})], \quad (5.77)$$

where the superscript C is used to denote the Coulomb potential, the sum is over the $\ell = 1$ lattice harmonics, and the force is in the coordinate system of the $Y_{\ell m}$, *i.e.* the rotated frame for a non-representative atom (Fig. 5.6).

The IBS contributions in Eqn. 5.76 are divided into core and valence terms. The core states in the LAPW method are fully relaxed in the radial direction, and so their IBS contribution to the force arises because of the spherical constraint (*n.b.* the IBS correction is zero for exact KS orbitals). Thus the core term is the force on the core charge due to the non-spherical crystal potential,

$$F^\alpha_{core} = \frac{1}{\delta\mathbf{r}_\alpha}[\sum_{\alpha,c}\delta\epsilon_{\alpha,c} - \int \mathrm{d}^3\mathbf{r}\ \rho_\alpha(\mathbf{r})\delta V_{KS}(\mathbf{r})]$$
$$= -\int \mathrm{d}^3\mathbf{r}\rho_\alpha(\mathbf{r})\nabla V_{KS}(\mathbf{r}), \quad (5.78)$$

where $\rho_\alpha(\mathbf{r})$ denotes the core charge density due to atom α. Both this term and F^α_{HF} depend on an accurate description of the $\ell = 1$ potential near the nucleus, and this depends on the $\ell = 1$ charge density. In practice, this means that force calculations require the use of a non-spherical mesh that extends much closer to the nucleus than that needed for total energy calculations alone.

The valence contribution to the IBS force, F^α_{IBS} has the form derived by Bendt and Zunger [16] with an extra term from the discontinuity at the sphere boundary.

$$F^\alpha_{IBS} = -\sum_{\mathbf{k},j} W(\mathbf{k},j)[< \frac{\delta\varphi_{\mathbf{k},j}}{\delta\mathbf{r}_\alpha}|H - \epsilon_{\mathbf{k},j}|\varphi_{\mathbf{k},j} > + < \varphi_{\mathbf{k},j}|H - \epsilon_{\mathbf{k},j}|\frac{\delta\varphi_{\mathbf{k},j}}{\delta\mathbf{r}_\alpha} > + \mathbf{D}^\alpha_{\mathbf{k},j}], \tag{5.79}$$

where the sum is over occupied states, the $\varphi_{\mathbf{k},j}$ are the wavefunctions and

$$\mathbf{D}^\alpha = \int\int [(\varphi^* T\varphi)|_{MT} - (\varphi^* T\varphi)|_I]\, \mathrm{d}^2\mathbf{S}_\alpha, \tag{5.80}$$

where the integral is over the surface of the sphere ($\mathrm{d}^2\mathbf{S}_\alpha$ denotes the unit surface normal), MT denotes use of the sphere representation, I denotes use of the planewave interstitial representation, and the band indices $(\mathbf{k}, j)$ are suppressed.

The change in the basis functions, $\phi_\mathbf{G}$, from moving atom α is zero outside the corresponding sphere, α, and inside it is given by

$$\frac{\delta\phi_\mathbf{G}}{\delta\mathbf{r}_\alpha} = i(\mathbf{k}+\mathbf{G})\phi_\mathbf{G} - \nabla\phi_\mathbf{G} + \sum_{\ell m} Y_{\ell m}[\frac{\delta(a_{\ell m}u_\ell)}{\delta\mathbf{r}_\alpha} + \frac{\delta(b_{\ell m}\dot{u}_\ell)}{\delta\mathbf{r}_\alpha}], \tag{5.81}$$

where the last term includes the contribution due to the variation of the u_ℓ and $\dot{u}_\ell$ that results from the change in the spherical component of the potential as the center of the sphere is moved. The effect of this term (the frozen augmentation correction) is small [211], and is neglected in the following. Thus, in terms of the basis functions,

$$\begin{aligned} F^\alpha_{IBS} &= -\sum \mathbf{k}, jW(\mathbf{k},j) \\ &\times \sum_{\mathbf{G},\mathbf{G}'} [i(\mathbf{G}' - \mathbf{G}c_\mathbf{G})^* c_{\mathbf{G}'} < \phi_\mathbf{G}|H - \epsilon_{\mathbf{k},j}|\phi_{\mathbf{G}'} >_\alpha \\ &- < \nabla\phi_\mathbf{G}|H - \epsilon_{\mathbf{k},j}|\phi_{\mathbf{G}'} >_\alpha - < \phi_\mathbf{G}|H - \epsilon_{\mathbf{k},j}|\nabla\phi_{\mathbf{G}'} >_\alpha + \mathbf{D}^\alpha_{\mathbf{k},j}], \end{aligned} \tag{5.82}$$

where $<>_\alpha$ denotes integrating over the sphere α only. This expression can be further simplified by combining the last three terms, grouping the kinetic and potential energy terms, and using the divergence theorem. In this way, one may obtain,

$$
\begin{aligned}
F^{\alpha}_{IBS} = & -\sum \mathbf{k}, jW(\mathbf{k}, j) \\
& \times \sum_{\mathbf{G},\mathbf{G}'} [i(\mathbf{G}' - \mathbf{G}) c^*_{\mathbf{G}} c_{\mathbf{G}'} < \phi_{\mathbf{G}} | H - \epsilon_{\mathbf{k},j} | \phi_{\mathbf{G}'} >_\alpha \qquad (5.83) \\
& - \int\!\!\int \phi^*_G (T - \epsilon_{\mathbf{k},j} \phi_{\mathbf{G}'} \mathrm{d}^2 \mathbf{S}_\alpha] + \int \mathrm{d}^3 \mathbf{r} V_{KS}(\mathbf{r}) \nabla \rho_v(\mathbf{r}),
\end{aligned}
$$

where the volume integrals are over the sphere α, the surface integral is with the interstitial representation, $\rho_v(\mathbf{r})$ is the valence part of the charge density and T is the kinetic energy operator, which on the surface of the sphere is non-relativistic. Although somewhat cumbersome, these expressions can be computed directly within an LAPW code (see Yu *et al.* [211]).

5.15 Density Functional Perturbation Theory and Linear Response

It is often of interest to calculate the responses of solids to perturbations at finite wavevector, $\mathbf{q}$. Examples include calculations of phonon frequencies, dielectric response, electron phonon couplings and magnetic susceptibility. If $\mathbf{q}$ is commensurate, this can in principle be done directly by using a supercell and applying the perturbation. However, because the cost of calculations rapidly increases with the size of the unit cell, this approach is often impractical, and, in any case, incommensurate $\mathbf{q}$'s are excluded. The alternative to this is provided by density functional perturbation theory (DFPT). Two equivalent formalisms were developed by Baroni and co-workers [8, 49] and Gonze [57, 58, 59]. The Gonze approach is based on perturbative expansion of the density functional energy functional, while the approach of Baroni and co-workers is based on self-consistent Green's function methods. Here a brief overview of DFPT is given, followed by aspects particular to the LAPW method. The formalism is for systems with a finite gap, *i.e.* insulators. In metals, there are corrections, which can be large, due to changes in occupation number at the Fermi surface [9, 51]. A detailed review of DFPT may be found in Ref. [9].

The starting point is expansion of Kohn-Sham orbitals, densities and self-consistent potentials in perturbation series,

$$X(\lambda) = X^{(0)} + \lambda X^{(1)} + \lambda^2 X^{(2)} + \lambda^3 X^{(3)} + \ldots , \qquad (5.84)$$

where $X(\lambda)$ is the quantity of interest and λ is the perturbing parameter. DFPT rests on the so-called $2n+1$ theorem. The statement is that variations of the energy up to order $2n+1$ in a perturbation are determined using only variations in Kohn-Sham orbitals up to order n. This is useful because connected with this, variations in the Kohn-Sham orbitals are given by solutions of Sternheimer type equations [188], and these can be built up to various orders. For example, considering atomic displacement as the perturbation, $E^{(0)}$ gives the total energy, the $E^{(1)}$ are the the atomic forces, the $E^{(2)}$ are the curvatures of the energy surface (sums of force constants), which determine the phonon spectrum, the $E^{(3)}$ are cubic anharmonic terms in the energy surface, and so on.

The $2n+1$ theorem says that the atomic forces can be determined directly from the unperturbed, $n=0$, Kohn-Sham orbitals. This corresponds to the Hellmann-Feynman theorem in density functional language. The linear response of the Kohn-Sham orbitals, $\varphi^{(1)}$, is then sufficient for both $E^{(2)}$ (phonons) and $E^{(3)}$. The Sternheimer equation that needs to be solved to obtain the $\varphi^{(1)}$ is

$$(H^{(0)}-\epsilon_i^{(0)})\varphi_i^{(1)}=-(H^{(1)}-\epsilon_i^{(1)})\varphi_i^{(0)}, \tag{5.85}$$

where i is the band index and $\epsilon_i^{(0)}$ and $\epsilon_i^{(1)}$ are the unperturbed, and lowest order, $n=1$, perturbed Kohn-Sham eigenvalues, respectively. In Eqn. 5.85, $H^{(0)}$ is the unperturbed Kohn-Sham Hamiltonian, and $H^{(1)}$ is its lowest order variation with λ, including not only the external perturbation but also contribution via the Hartree and exchange correlation potential arising from variations in the charge density due to the perturbation.

$$H^{(1)}(\mathbf{r})=V_{ext}^{(1)}(\mathbf{r})+e^2\int d^3\mathbf{r}'\,\frac{\rho^{(1)}(\mathbf{r}')}{|\mathbf{r}-\mathbf{r}'|}+\int d^3\mathbf{r}'\,\frac{\delta V_{xc}}{\delta n(\mathbf{r}')}\rho^{(1)}(\mathbf{r}'), \tag{5.86}$$

where $\rho^{(1)}$ is the $n=1$ variation in the density and it is assumed that we are working with a density functional formalism where the kinetic energy operator does not depend on the external perturbation (otherwise there is a contribution $T^{(1)}$ in Eqn. 5.86). The key point is that $H^{(1)}$ depends on $\rho^{(1)}$, and therefore Eqns. 5.85 and 5.86 must be solved self-consistently.

The normalization condition, $<\varphi_i|\varphi_i>=1$ leads to the constraint ,

$$<\varphi_i^{(0)}|\varphi_i^{(1)}>+<\varphi_i^{(1)}|\varphi_i^{(0)}>=0, \tag{5.87}$$

but recognizing the phase freedom in the Kohn-Sham orbitals, it is convenient to use the stronger constraint,

$$<\varphi_i^{(0)}|\varphi_i^{(1)}>=0, \tag{5.88}$$

which leads to an expression for the $n = 1$ variation in the Kohn-Sham eigenvalues,

$$\epsilon_i^{(1)} =<\varphi_i^{(0)}|H^{(1)}|\varphi_i^{(0)}>. \tag{5.89}$$

Finally, the $n = 1$ variation in the density is

$$\rho^{(1)}(\mathbf{r}) = \sum_{occ}[\varphi_i^{(1)*}(\mathbf{r})\varphi_i^{(0)}(\mathbf{r}) + \varphi_i^{(0)*}(\mathbf{r})\varphi_i^{(1)}(\mathbf{r})]. \tag{5.90}$$

Given H^0, $H^{(1)}$, and the $\varphi_i^{(0)}$, Eqn. 5.85 can be reduced to a simple linear problem in the basis set of interest, using Eqn. 5.89 for $\epsilon_i^{(1)}$. Thus the linear response problem can be solved as follows. First the self consistent Kohn-Sham problem is solved for $H^{(0)}$ yielding, $\rho^{(0)}$, the $\varphi_i^{(0)}$ and the $\epsilon_i^{(0)}$. Then, keeping these fixed, Eqns. 5.85, 5.90 and 5.86 are iterated to find the self-consistent $H^{(1)}$, and $\varphi_i^{(1)}$, $\rho^{(1)}$. Linear response calculations have been implemented for phonons and other perturbations in a variety of planewave codes, such as `abinit` [60] and `pwscf` [10].

LAPW linear response calculations were implemented in 1994 by Yu and Krakauer [212] and were demonstrated by calculations for several oxides [213, 200, 201] and other materials [199]. This implementation was considerably more complicated than in a planewave method because of the incomplete basis set corrections, and kinetic energy contribution on the sphere boundary, as discussed in the section on LAPW force calculations, above, and because of the special character of the LAPW radial functions. In particular, the LAPW basis inside the spheres made of functions constructed to be solutions of radial equations, based on $H^{(0)}$ near the band energy. However, depending on the perturbation, the solutions $\varphi_i^{(1)}$ of the Sternheimer equation, 5.85 are very different, and cannot generally be reasonably represented by the LAPW $u_\ell Y_{\ell m}$ and $\dot{u}_\ell Y_{\ell m}$. For example, for phonons, the perturbation is a displacement wave,

$$\mathbf{w}_\alpha(\mathbf{R}) = \mathbf{w}_{0,\alpha}\, e^{i\mathbf{q}\cdot\mathbf{R}}, \tag{5.91}$$

where $\mathbf{q}$ is the wavevector, α is an atom label, and $\mathbf{R}$ is a lattice vector. Thus,

$$V_{ext}^{(1)}(\mathbf{r}) = -\sum_{\mathbf{R},\alpha}\frac{Z_\alpha \mathbf{w}_\alpha(\mathbf{R})\cdot(\mathbf{r}-\mathbf{R}-\mathbf{R}_\alpha)}{|\mathbf{r}-\mathbf{R}-\mathbf{R}_\alpha|^3}, \tag{5.92}$$

where $\mathbf{R}_\alpha$ is the position of atom α in the unit cell. Yu and Krakauer [212] proceeded by noting that the LAPW basis functions are a very good basis for solving the electronic structure problem for atoms located at the centers of the LAPW spheres. Thus,

$$\varphi_i^{(1)} = \sum_{\mathbf{G}} [c_{\mathbf{G},i}^{(1)} \phi_{\mathbf{G}} + c_{\mathbf{G},i} \phi_{\mathbf{G}}^{(1)}], \tag{5.93}$$

where i denotes the band index and $\mathbf{k}$ point, the c_i are the variational coefficients and the ϕ_i are the LAPW basis functions. The LAPW basis functions are planewaves in the interstitial, where they do not depend on the atomic coordinates. Therefore, the $\phi_i^{(1)}$ are non-zero only inside the spheres corresponding to the atoms that are moved in the displacement wave. The variation, $\phi_i^{(1)}$, contains two types of terms: terms due to the shift of the sphere and terms due to the relaxation of u and $\dot{u}$ due to the change in the spherical potential. As in the case of the forces, discussed above, the second class of terms is small and so the frozen augmentation approximation can be made. This approximation can be improved if needed by the use of local orbital extensions to the basis. Within this frozen augmentation approximation,

$$\phi_{\mathbf{G}}^{(1)} = \frac{\partial \phi_{\mathbf{G}}}{\partial \mathbf{w}} \cdot \mathbf{w} = [i(\mathbf{k}+\mathbf{G}) - \nabla]\phi_{\mathbf{G}} \cdot \mathbf{w}, \tag{5.94}$$

where the subscripts on $\mathbf{w}$ are suppressed. This then is used to construct the Sternheimer-like equation for the $c^{(1)}$ in terms of the unperturbed solution, and $V^{(1)}$ (and thus implicitly $\rho^{(1)}$), which can then be solved self-consistently.

The key points are (1) kinetic energy terms arising from discontinuities in derivatives at the sphere boundaries need to be explicitly considered, and (2) the appropriate basis set should be used for $\varphi^{(1)}$, *i.e.* including the change in the LAPW basis functions under $V^{(1)}$, if this is not already included in the variational flexibility provided by u and $\dot{u}$ (and the additional u if local orbitals are used). While Yu and Krakauer implemented the linear response calculation for phonons and effective charges, these considerations are generic to density functional perturbation theory calculations with the LAPW method.

5.16 Second Variational Treatment of Spin-Orbit Effects

The spin-orbit term (neglected in the scalar relativistic approximation) is important for the band structures and other properties of materials containing heavier elements, as well as for some properties (*e.g.* the magneto-crystalline anisotropy) of lighter magnetic materials. Unfortunately, the spin-orbit term couples spin-up and spin-down wavefunctions. This means that if one began (without spin-orbit) with n basis functions, and therefore an $n \times n$ secular equation for each spin, and then included spin-orbit in the Hamiltonian, one would end up with a $2n \times 2n$ secular equation to solve at each **k**-point. Because of the cubic scaling of the computing time with the size of the secular equation, this means that a calculation including spin-orbit would require roughly eight times the resources of a single spin scalar relativistic calculation. This difficulty may, however, be avoided by recognizing that spin-orbit is generally a small effect. This fact may be exploited by diagonalizing the Hamiltonian including spin-orbit, in the space of the orbitals for the low lying bands as obtained in a scalar relativistic step [113]. As discussed below, the second variational approach does involve approximations, which may be of importance in *e.g.* actinides or heavy p elements. As such, it is better to use a direct treatment of spin-orbit as part of the usual Hamiltonian in non-collinear magnetic calculations, since the matrix size is already $2n \times 2n$ in that case.

The second variational method for relativistic calculations is derived from the scalar relativistic procedure, discussed above, as illustrated in Fig. 5.16. First, the scalar relativistic bands are calculated by setting up and diagonalizing the secular equations, exactly as if spin-orbit were not to be included. Next, a second variational secular equation is set up, using, as basis functions, the lowest N scalar relativistic orbitals (for both spins) as calculated in the previous step. This yields a $2N \times 2N$ system as opposed to the $2n \times 2n$ system that would be obtained using the usual LAPW basis functions. Since in LAPW calculations, the number of bands of interest is almost always much smaller than the number of basis functions, N can be chosen much smaller than n, if spin-orbit is indeed a weak effect. Further, in the second variational basis, the overlap, S, is diagonal, and the Hamiltonian is just the spin-orbit term plus the scalar relativistic eigenvalues on the diagonal. The spin-orbit contribution to the Hamiltonian is also as straight-

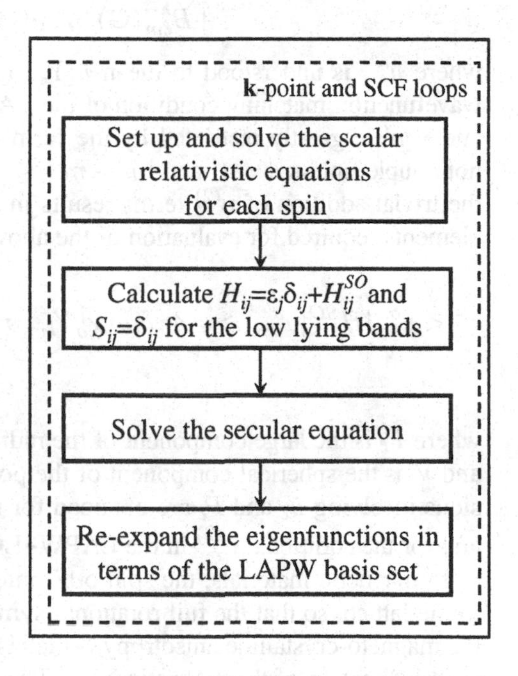

Figure 5.16. The second variational procedure for treating spin orbit. i and j are indices that run over spins and the bands to be included in the second variation.

forward to calculate in the second variational basis as it is in the original LAPW basis. It is negligible outside the spheres, while inside them it can to a very good approximation be calculated as if the atoms were spherical (*n.b.* spin-orbit weights heavily the region near the nucleus, where a spherical approximation is best). Thus, spin-orbit matrix elements between the various radial functions inside a sphere can be precalculated; then the second variational spin-orbit matrix elements from a given atom are just linear combinations of these, fixed by the $A_{\ell m}$ and $B_{\ell m}$.

$$\begin{aligned} < \varphi^{\sigma}_{\mathbf{G}}|H^{SO}|\varphi^{\sigma'}_{\mathbf{G}'} > = \sum_{\ell m,\ell' m'} & [A^{*}_{\ell m}(\mathbf{G})A_{\ell' m'}(\mathbf{G}') < u^{\sigma}_{\ell m}|H^{SO}|u^{\sigma'}_{\ell' m'} > \\ & + B^{*}_{\ell m}(\mathbf{G})A_{\ell' m'}(\mathbf{G}') < \dot{u}^{\sigma}_{\ell m}|H^{SO}|u^{\sigma'}_{\ell' m'} > \\ & + A^{*}_{\ell m}(\mathbf{G})B_{\ell' m'}(\mathbf{G}') < u^{\sigma}_{\ell m}|H^{SO}|\dot{u}^{\sigma'}_{\ell' m'} > \\ & + B^{*}_{\ell m}(\mathbf{G})B_{\ell' m'}(\mathbf{G}') < \dot{u}^{\sigma}_{\ell m}|H^{SO}|\dot{u}^{\sigma'}_{\ell' m'} >], \end{aligned} \tag{5.95}$$

where $u^{\sigma}_{\ell m}$ is understood to mean $u_{\ell}Y_{\ell m}\chi_{\sigma}$, the $A_{\ell m}$ and $B_{\ell m}$ are from the wavefunction matching condition of the LAPW method, and the sum over ℓm and $\ell' m'$ is greatly restricted by the form of H^{SO}, which in particular does not couple unless $\ell = \ell'$ and $|m - m'| \leq 1$. A very similar expression, with the trivial addition of $u^{(2)}$ terms results in the LAPW+LO basis. The matrix elements required for evaluation of the above expression are

$$\begin{aligned} < u^{\sigma}_{\ell m}|H^{SO}|u^{\sigma'}_{\ell' m'} > = 4\pi\delta_{\ell\ell'}(\chi^{\dagger}_{\sigma}Y^{*}_{\ell m}\,\sigma\cdot\mathbf{L}\,Y_{\ell' m'}\chi_{\sigma'}) \\ \times \int P_{\ell}P_{\ell'}(\frac{1}{2Mc})^{2}\frac{1}{r}\frac{\mathrm{d}V}{\mathrm{d}r}\mathrm{d}r, \end{aligned} \tag{5.96}$$

where P_{ℓ} is the large component of the radial function, u_{ℓ} (Eqns. 5.28 - 5.32), and V is the spherical component of the potential. Exactly analogous expressions involving $\dot{u}_{\ell}$ and $\dot{P}_{\ell}$ are obtained for the other three terms in Eqn. 5.95, and for the additional $u^{(2)}_{\ell}$ in the LAPW+LO case.

In magnetic materials, the spin orbit interaction couples the magnetization to the lattice, so that the full rotational symmetry is lost. This is the origin of the magneto-crystalline anisotropy, which is the dependence of the total energy on the direction of the magnetization. The consequence that needs to be considered in calculations is that the symmetry that can be used in calculations is at most a uniaxial symmetry, with the axis along the magnetization. Inversion symmetry (P) is also lost, but inversion times time reversal (PT) is maintained. This needs to be considered when analyzing wavefunctions at **k** and -**k** in non-centrosymmetric crystals. This is because the time reversal operation reverses magnetic fields, as discussed by Kunes and co-workers [95]. The second variational approach normally converges well with respect to the number of bands included, in cases where it is effective. However, in practice, it is important to check this convergence.

5.17 Spin-Orbit with $p_{1/2}$ Local Orbitals

For the reasons discussed above, this second variational approach is much more efficient than approaches, like the relativistic APW (RAPW) method, which treat the spin-orbit interaction on the same level as the rest of the Hamiltonian [111]. Unfortunately, the method, as described above, is not without deficiencies. MacDonald *et al.* report tests in which they compare the results of the second variational approach (SO-LAPW, in their notation) with the more reliable RAPW method. For Pd metal they find essentially the same band structure with the two methods, but in Pb significant differences were found for states with large $p_{1/2}$ components. They showed that these differences did not result from the reduction in the flexibility of the basis that occurs when all but the low lying bands are discarded prior to the second variational step, but rather were due to limitations inherent in the original LAPW basis. In particular, the scalar relativistic (j averaged) radial functions u_ℓ and $\dot{u}_\ell$ are not sufficiently flexible to represent the behavior of the $p_{1/2}$ orbital. Unlike the $p_{3/2}$ and higher ℓ orbitals, the $p_{1/2}$ orbital has a finite amplitude at the nucleus. This, however, cannot be reproduced by scalar relativistic radial functions, which for $\ell = 1$ vanish at $r = 0$. Further, related to the non-vanishing amplitude at $r = 0$, relativistic effects are stronger for the $p_{1/2}$ orbital, which shows significant contraction relative to the $p_{3/2}$ and scalar relativistic orbitals. Nordström and co-workers presented a critical examination of the effect of this problem for light actinides [134].

In a nutshell the problem is this: The basis inside the spheres is constructed to reproduce scalar relativistic wavefunctions; this is then being used to solve variationally the Dirac equation. If the Dirac wavefunctions differ significantly from the scalar relativistic wavefunctions, this procedure will break down, and this occurs when the p states of heavy elements are involved, *i.e.* calculations involving the 6p elements beginning with Hg and actinides.

Because of the computational advantages of the SO-LAPW approach over the RAPW method (the secular equations are half the size for a collinear magnetic system), it would be quite desirable to have a method of improving it for 6p and other materials with heavy atoms. Clearly, the essential ingredient is to improve the variational freedom for $\ell = 1$ orbitals inside the spheres. One avenue is to include in the second variational step, along with the low lying band states, $\ell = 1$ local orbitals constructed using a $j = 1/2$ radial function along with the usual u_ℓ and $\dot{u}_\ell$.

Following this suggestion, made in the first edition of this book, Kunes and co-workers, implemented the method and presented test calculations for Th metal demonstrating its efficacy [96]. These tests were extended to a large number of semiconductors by Carrier and Wei [31], who showed that the method captures essentially all of the spin-orbit interaction effects compared with full relativistic treatments. In the implementation of Kunes and co-workers, the local orbital that is added is not orthogonal to the LAPW functions of the first variational step, so a generalized eigenvalue problem must be solved in the second variational, spin orbit step. This, however, leads to only a small overhead, because most of the computational time is spent in the first variational step.

5.18 Iterative Diagonalization

As discussed in Chapter 3, iterative diagonalization is very advantageous in planewave methods because it allows one to avoid diagonalization of very large secular equations, of which only a small fraction of the eigenstates are of interest (*i.e.* occupied or at least near the Fermi energy). This is also true in the LAPW method, though with a normally smaller ratio between the number of basis functions (size of the secular equation) and the number of occupied bands. The DIIS method [210] was implemented in the context of the LAPW method in 1989 [165]. This resulted in a large reduction in the computational cost of calculations, but also revealed some problems that are not present in planewave calculations. Essentially, this method involves the use of a variational step in the space of the eigenvectors from the previous step and preconditioned residuals of these eigenvectors (see Chapter 3).

The method converges rapidly if the preconditioned residuals nearly span the space of the difference between the trial eigenvectors from the previous step and the true eigenvectors. In planewave methods, this is generally the case with simple preconditioning schemes based on the diagonal elements of the Hamiltonian matrix. However, in the LAPW method, it was empirically found that the simple preconditioning schemes are not as effective, and this problem is worse in the APW+LO method, which has more compact, less diagonally dominant secular equations. At present, the problem of finding an effective and computationally efficient preconditioning for the residuals in the context of the LAPW and APW+LO methods is not solved. It is frequently observed that after some number of DIIS based self-consistent iterations, the convergence, which can be measured by the norm of the residuals, slows down. At this point, the heuristic of using an exact diagonalization step needs to be employed. The result is that DIIS based iterative diagonalization is effective in reducing the time required for LAPW calculations, but because of the needed exact diagonalization steps every few iterations, not nearly as effective as it would be otherwise.

Chapter 6

CAR-PARRINELLO AND THE LAPW METHOD

6.1 Preliminaries

The development of the Car-Parrinello (CP) technique and its application to planewave based calculations resulted in a tremendous increase in the size and complexity of the problems that can be treated within a fully ab initio LDA framework. This and related techniques combined with ultrasoft pseudopotentials quickly enabled detailed structural investigations for unit cells of several hundred atoms [189, 24]. However, besides the general observations of Car and Parrinello, there are several special features of planewaves that are exploited, and these are crucial in making the CP method as efficient as it is (see Chapter 3). These are difficult to generalize to non-planewave basis sets, and because of this CP-like algorithms have not yet had as large an impact with non-planewave basis methods. Instead, most LAPW and other non-planewave basis methods have continued to use conventional algorithms, while localized basis set methods aimed at large systems have been developed using both conventional, diagonalization based algorithms and order N approaches, especially based on density matrix approaches [135, 106, 56, 184, 89].

The LAPW method, however, has strong similarities to planewave pseudopotential approaches. In particular, the wavefunctions have the same labeling, and their long range characters are the same as well. In particular, the modifications to the planewaves are due to the presence of atoms, and the range over which each atom modifies the planewaves is strictly confined to a region around that atom. Considering the importance of the CP approach and the connections between the LAPW method and planewave methods, especially ultrasoft pseudopotential planewave methods, it is worthwhile to develop these connections further and explore the possibility of using CP algorithms in LAPW codes. We note that there has been progress along these lines, and most notably, the PAW method mixes ideas from both approaches with considerable success [22].

Both planewave pseudopotential and LAPW methods begin with planewaves, and are then forced to confront the fact that near the nuclei the potential is strong and therefore the valence wavefunctions are strongly varying. Accordingly, modifications are needed in order to avoid the extremely slow convergence with basis set size that results. In the planewave pseudopotential approach, the Hamiltonian, H, is replaced by a pseudo-Hamiltonian, $\tilde{H}$, that is constructed to reproduce the original eigenvalue spectrum in the valence region as well as the wavefunctions outside some distance r_c of the nucleus. As discussed in Chapter 3, more rapid convergence is obtained using "ultrasoft" pseudopotentials at the expense of modifying the (initially diagonal) overlap as well. Thus the matrix elements occurring in the secular equation are:

$$H^{PP}_{\mathbf{GG'}} = < \phi_{\mathbf{k+G'}} | \tilde{H} | \phi_{\mathbf{k+G}} >, \tag{6.1}$$

and for ultrasoft pseudopotentials ,

$$S^{PP}_{\mathbf{GG'}} = < \phi_{\mathbf{k+G'}} | \tilde{S} | \phi_{\mathbf{k+G}} >, \tag{6.2}$$

where ϕ is a planewave, and, as mentioned, orthogonality to the core and the strong atomic potential inside r_c are handled via $\tilde{H}$ and $\tilde{S}$.

The LAPW method approaches the problem in an apparently different fashion. The true all-electron Hamiltonian and the standard overlap are retained, but the planewaves, $\phi_{\mathbf{k+G}}$ are modified and become the LAPWs, $\tilde{\phi}_{\mathbf{k+G}}$ within the spheres.

$$H^{LAPW}_{\mathbf{GG'}} = < \tilde{\phi}_{\mathbf{k+G'}} | H | \tilde{\phi}_{\mathbf{k+G}} >, \tag{6.3}$$

and

$$S^{LAPW}_{\mathbf{GG'}} = < \tilde{\phi}_{\mathbf{k+G'}} | \tilde{\phi}_{\mathbf{k+G}} >, \tag{6.4}$$

As in the pseudopotential case, the valence eigenvalue spectrum is retained, and rapid convergence with respect to the number of basis functions is achieved. In both pseudopotential and LAPW methods, a similar price is to be paid: the method is only valid within a certain energy range, (*e.g.* the core states are cutoff). It is the isomorphism of Eqns. 6.1 and 6.2 with Eqns. 6.3 and 6.4, along with the fact that, in both cases, the modification is local to an atom, that is the heart of the connections between the methods and suggests the possible applicability of CP-like algorithms to the LAPW method.

Not surprisingly, connections between the LAPW method and planewave basis pseudopotential approaches have been noted in the past and exploited in certain contexts. In fact the subject has a long history going back to the recognition that the (not linearized) APW method has the same isomorphism at a fixed energy and that this is useful because in many cases only one energy, *i.e.* the Fermi energy, E_F, is of interest (recall that the original Phillips-Kleinman pseudopotential is also energy dependent, as discussed in Chapter 3). Accordingly, connections between the APW method and pseudopotential methods had already been elucidated in the early 1970's [214, 70, 103] More recently, Goedecker and Maschke [53] explicitly transformed the LAPW method into a planewave method with a separable pseudopotential involving with an overlap contribution and suggested that the use of this transformation could speed up calculations. Blöchl [21] has discussed a related transformation. Singh and co-workers [170, 109] transformed the LAPW method into an equivalent real space [194] pseudopotential method, also with an overlap component.

The purpose of this chapter is to elucidate these relationships, and to suggest directions for incorporation of CP-like algorithms into LAPW calculations. Although some work in this direction has already been done, we think that it is an area where much remains to be done.

6.2 The Transformation of Goedecker and Maschke

The transformation of Goedecker and Maschke [52, 53] uses an operator technique to perform the matching of the basis functions at the LAPW sphere boundaries. To simplify the expressions, radial functions consisting of linear combinations of the standard u_ℓ and $\dot{u}_\ell$ are used to augment the planewaves. These are R_ℓ^1 and R_ℓ^2, which are determined by,

$$R_\ell^1(R_\alpha) = 1, \qquad \frac{\partial R_\ell^1(R_\alpha)}{\partial r} = 0, \tag{6.5}$$

and

$$R_\ell^2(R_\alpha) = 0, \qquad \frac{\partial R_\ell^2(R_\alpha)}{\partial r} = 1, \tag{6.6}$$

where R_α is the sphere radius. In terms of these, the matching coefficients (which were $a_{\ell m}$ and $b_{\ell m}$ with the standard u_ℓ and $\dot{u}_\ell$ radial functions) become $\alpha_{\ell m}$ and $\beta_{\ell m}$ and are given by averages over the sphere boundary,

$$\alpha_{\ell m} = \int \mathrm{d}^2\mathbf{S}\; Y_{\ell m}^*(\mathbf{r})\phi(\mathbf{r}), \tag{6.7}$$

and

$$\beta_{\ell m} = \int d^2\mathbf{S}\, Y^*_{\ell m}(\mathbf{r})\hat{\mathbf{n}} \cdot \nabla\phi(\mathbf{r}), \tag{6.8}$$

where ϕ is the interstitial representation of the basis function, $\hat{\mathbf{n}}$ is the unit normal, and the atom is taken to be at the origin. Goedecker and Maschke evaluate the $\alpha_{\ell m}$ and $\beta_{\ell m}$ in terms of the projectors,

$$p^{1*}_{\ell m}(\mathbf{G}) = \frac{1}{\Omega^{1/2}} \int d^2\mathbf{S}\, Y^*_{\ell m}(\mathbf{r}) e^{i\mathbf{G}\cdot\mathbf{r}} = \frac{4\pi\, i^\ell}{\Omega^{1/2}} j_\ell(|\mathbf{G}|R_\alpha) Y^*_{\ell m}(\mathbf{G}), \tag{6.9}$$

and

$$\begin{aligned} p^{2*}_{\ell m}(\mathbf{G}) &= \frac{1}{\Omega^{1/2}} \int d^2\mathbf{S}\, Y^*_{\ell m}(\mathbf{r})\, \hat{\mathbf{n}} \cdot \nabla e^{i\mathbf{G}\cdot\mathbf{r}} \\ &= \frac{4\pi i^\ell}{(2\ell+1)\Omega^{1/2}} [\ell j_{\ell-1}(|\mathbf{G}|R_\alpha) - (\ell-1) j_{\ell+1}(|\mathbf{G}|R_\alpha)] Y^*_{\ell m}(\mathbf{G}), \end{aligned} \tag{6.10}$$

where the extra factors of $1/\Omega^{1/2}$ relative to the papers of Goedecker and Maschke are due to the different normalization of the interstitial planewaves that is used in this book.

Thus, in terms of the coefficients, $c_{\mathbf{k}+\mathbf{G}}$ of the interstitial planewave representation of the wavefunction, φ, the sphere coefficients ($\bar{\alpha}_{\ell m}$ and $\bar{\beta}_{\ell m}$) needed for the charge density (corresponding to the $A_{\ell m}$ and $B_{\ell m}$ of Fig. 5.11) are

$$\bar{\alpha}_{\ell m} = \sum_{\mathbf{G}} p^{1*}_{\ell m}(\mathbf{k}+\mathbf{G}) c_{\mathbf{k}+\mathbf{G}}, \tag{6.11}$$

and

$$\bar{\beta}_{\ell m} = \sum_{\mathbf{G}} p^{2*}_{\ell m}(\mathbf{k}+\mathbf{G}) c_{\mathbf{k}+\mathbf{G}}. \tag{6.12}$$

Goedecker and Maschke write the Hamiltonian matrix elements by beginning with an empty lattice containing no spheres. They then add correction terms for (1) the removal of the planewave kinetic energy in the spheres, (2) the interstitial potential, and (3) the contribution due to the augmenting functions inside the spheres. Thus in Rydberg units,

$$\begin{aligned} H_{\mathbf{G}\mathbf{G}'} = (\mathbf{k}+\mathbf{G})^2 \delta_{\mathbf{G},\mathbf{G}'} &- (\mathbf{k}+\mathbf{G}) F_{(\mathbf{G}-\mathbf{G}')} (\mathbf{k}+\mathbf{G}') \\ &+ \frac{1}{\Omega} \int_I d^3\mathbf{r} e^{i(\mathbf{G}'-\mathbf{G})\cdot\mathbf{r}} \\ &+ \sum_{\ell m} \sum_{\ell' m'} \sum_{\kappa\kappa'} p^{\kappa}_{\ell m}(\mathbf{k}+\mathbf{G}) H^{\kappa\kappa'}_{\ell m,\ell' m'} p^{\kappa' *}_{\ell' m'}(\mathbf{k}+\mathbf{G}'), \end{aligned} \tag{6.13}$$

where one atom at the origin is assumed (otherwise the second and fourth terms are to be summed over atoms), κ and κ' take the values 1 and 2, the potential integral is over the interstitial, the $H^{\kappa\kappa'}_{\ell m,\ell' m'}$ are the matrix elements of the radial functions (*n.b.* the kinetic energy here has been written as $(\nabla\varphi^* \cdot \nabla\varphi)$ to avoid discontinuities at the sphere boundary that contribute to the forces; this is the form of the kinetic energy used in the APW and APW+LO methods, but not in most LAPW codes, which often use $\varphi^*\nabla^2\varphi$). $F_{\mathbf{q}}$ is given by

$$F_{\mathbf{q}} = \frac{1}{\Omega}\int_{\alpha} \mathrm{d}^3\mathbf{r} e^{i\mathbf{q}\cdot\mathbf{r}}, \tag{6.14}$$

where the integral is over the sphere, α.

The overlap matrix elements are obtained in a similar way beginning with planewaves and (1) subtracting the planewave contribution from inside the spheres and (2) adding that due to the augmenting functions.

$$S_{\mathbf{G}\mathbf{G}'} = \delta_{\mathbf{G},\mathbf{G}'} - F_{(\mathbf{G}-\mathbf{G}')} + \sum_{\ell m}\sum_{\kappa\kappa'} p^{\kappa}_{\ell m}(\mathbf{k}+\mathbf{G}) S^{\kappa\kappa'}_{\ell m} p^{\kappa' *}_{\ell' m'}(\mathbf{k}+\mathbf{G}'), \tag{6.15}$$

where $S^{\kappa\kappa'}_{\ell m}$ is diagonal in ℓm because of the orthogonality of the $Y_{\ell m}$.

So what has been gained from this transformation? To begin with, it is apparent that the Hamiltonian matrix elements (Eqn. 6.13) consist of three contributions, (1) a standard kinetic energy (the first term), (2) a local pseudopotential (the third term) and (3) a non-local but separable pseudopotential (the second and fourth terms). Similarly the overlap involves only local and separable terms. These are exactly the type of terms that occur in planewave pseudopotential methods, for which CP methods are used; this would then seem to open the door for direct application of CP algorithms to the LAPW method. Moreover, the above transformation is exact; the resulting method, although appearing like a pseudopotential method, is precisely the LAPW method.

There are, however, some significant differences, both positive and negative. Two significant pluses relative to planewave pseudopotential methods are the very high accuracy of the LAPW method for any atom in the periodic table (*i.e.* no frozen core approximations, no shape approximations to the potential, even within the sphere radius etc.) and the extreme softness of the LAPW pseudopotential. On the negative side, this "pseudopotential" is extremely non-transferable. In other words, both the projectors and the matrix elements among the local orbitals are strongly dependent on the crystal potential, and because of this they must be recalculated at each step along the path to self-consistency. However, in a large scale calculation (many atoms per unit cell), the redetermination of this "pseudopotential" is a small part of the total computational work.

There are some more serious obstacles to the implementation of a CP-like LAPW code using this transformation. First of all, it is known that iterative diagonalization methods are less effective in the LAPW method than in planewave methods. This is because the secular equation is not as well preconditioned. This problem is even more severe in the more compact APW+LO basis. Secondly, there is the issue of the number of projectors appearing in Eqns. 6.13 and 6.15. A conventional separable pseudopotential has only a handful of projectors to evaluate, in addition to the local component. Even so, the operation of these is often the rate limiting step in CP codes using separable pseudopotentials. In the LAPW method, however, relatively high values of ℓ (typically $\ell_{max} = 8$) need to be included in order to have a reasonable matching between the sphere and interstitial representations of the wavefunctions. This results in a much larger number of projectors: $2(\ell_{max} + 1)^2$ *i.e.* 162 for $\ell_{max} = 8$. However, there is some hope that the transformation of Goedecker and Maschke may be useful in practice in spite of this. First of all, as they suggest, it may be possible to reduce ℓ_{max} somewhat by using smaller sphere radii than are common in standard LAPW calculations. Of course, this can only be carried so far. Smaller sphere radii require higher planewave cutoffs and more basis functions, so that some of the benefits of using the LAPW method rather than a conventional pseudopotential are lost as the radii are reduced. Secondly, and perhaps more significantly, King-Smith *et al.* [81] have developed a real space technique for operating separable pseudopotentials that could be applied to the Goedecker-Maschke formulation of the LAPW method. However, as far as we know, this has yet to be done.

Finally, it is interesting to note the similarities of the LAPW pseudopotential with the ultrasoft Vanderbilt pseudopotentials (Chapter 3). First of all, in both approaches the potential is equal to the all-electron potential in the interstitial as is the overlap (*i.e.* no off-diagonal contribution in this region). Secondly, in both cases there are separable non-local contributions from each atom and these have similar forms. Thirdly, inside the spheres (r_c in the case of the Vanderbilt pseudopotential) both the overlap and Hamiltonian are modified and the charge density is not equal to the square of the planewave part of the wavefunction but rather is augmented. Finally, the generation scheme has similarities to the construction of the augmentation in the LAPW method; in particular it is based on solutions, $u_\ell(r)$ of the radial Schrodinger equation at selected reference energies, and their matrix elements.

In the LAPW method, one conventionally regards the wavefunctions as being augmented, and the charge density as being computed directly from the $\varphi^*\varphi$, while with Vanderbilt pseudopotentials it is conventional to regard the wavefunctions as being just the planewave expansion and the charge density as being augmented. However, with the transformation of Goedecker and Mascke (and also that of Singh *et al.*; see below), it is apparent that one may regard the

LAPW wavefunctions as being just the planewave expansion, since this is what the effective Hamiltonian operates on. The distinction is simply one of semantics, *i.e.* to what component (wavefunction or Hamiltonian) various terms are assigned. There are, however, some significant differences between the two approaches. Favoring the Vanderbilt scheme is that fact that the LAPW method relies on matching conditions for the wavefunction at the sphere boundary; this is numerically less stable than the operations required in calculations with ultrasoft pseudopotentials. On the other hand, the LAPW method offers certain advantages. In particular, the entire LAPW pseudopotential is based on the true self-consistent crystal potential. Because of this, it may be expected to be more reliable, since it is in no way connected with any fixed atomic reference.

6.3 The Transformation of Singh *et al.*

Singh *et al.* [170, 109], following Goedecker and Maschke, also transformed the LAPW method to a pseudopotential-like form with a view to exploiting CP algorithms. However, instead of performing the transformations in reciprocal space, Singh *et al.* work directly in real space, using a specially constructed interpolation technique. This technique, called the projector basis method, is quite general, and has been used to operate semi-local pseudopotentials on planewaves and to construct CP-like mixed basis methods as well.

The motivation for working directly in real space is quite simple: The CP method with local pseudopotentials benefits greatly from the facts that the kinetic energy is diagonal in reciprocal space while the potential is diagonal in real space and that the planewaves can be efficiently transformed from one representation to another using FFTs. In the LAPW and most other cases involving planewaves (mixed basis, non-local pseudopotential etc.) the required operations are not diagonal in either real or reciprocal space. However, they can often be divided into operations that are diagonal in reciprocal space and operations that are local in real space. The significance of this is that while the number of planewaves scales with the number of atoms in the unit cell, the density of mesh points in real space does not, and therefore the number of points in the local vicinity of an atom is independent of the system size. Thus it may be anticipated that efficient methods for large systems will result if all operations that are non-diagonal in reciprocal space can be transformed to local real space operations. In other words having a large number of projectors per atom is a less serious problem in a real space formulation than in a reciprocal space formulation, provided that the number of projectors per atom does not increase with the system size.

The Projector Basis Method

The projector basis method [170, 109, 172] is basically an interpolation with a twist. The idea is to take the values of the wavefunction on the real space mesh points within some radius (for the LAPW method the sphere radius) of the nucleus, and to construct a smooth function that reproduces these values and has a form convenient for computing the required operations. The twist is to recognize that the goal is not to operate on arbitrary functions, but only to reproduce matrix elements involving wavefunctions that have zero Fourier components above some planewave cutoff G_{max}. This allows the construction of explicitly Hermetian operators (essential for most iterative diagonalization schemes) and the operation of potentials that are too strongly varying to be represented on the discrete FFT mesh. This is similar to the handling of the step function in the standard LAPW method, as discussed in Chapter 5.

In order to elucidate the method, it is convenient to consider first the simpler case of a conventional planewave calculation with a standard semi-local (*i.e.* ℓ dependent) pseudopotential. In this case, the goal is to couple the planewaves to the nonlocal component of the pseudopotential in real space.

To proceed, a set of projector basis functions, $f_j(\mathbf{r})$, centered at the nucleus is defined for each atom. Here j is an index labeling the functions and the atom index is suppressed. The explicit form of the $f_j(\mathbf{r})$ is not crucial, but it is convenient to use polynomials up to a cutoff power, p_{max} in a spherical harmonic representation, *i.e.*

$$f_j(\mathbf{r}) = r^p Y_{LM}(\mathbf{r}), \tag{6.16}$$

with $L = 0, 1, ..., p_{max}$, $M = -L, ..., L$, $p = L, ..., p_{max}$ and $L + p$ even. Here it is also convenient to use the real spherical harmonics to avoid complex arithmetic. With this choice, there are $n_j = (p_{max}+1)(p_{max}+2)(p_{max}+3)/6$ projector basis functions. This is a flexible basis for representing the planewave wavefunction provided that the number of $f_j(\mathbf{r})$ is comparable to the number of mesh points within the radius of interest, r_c (recall that the planewaves can be reexpanded into products of spherical Bessel functions and Y_{LM} and that near the nucleus a Taylor expansion for the Bessel functions is valid).

Next a matrix, $\mathbf{A}$, is defined that transforms from the values of the wavefunction on the local FFT mesh points (*i.e.* the points that lie inside the sphere, radius r_c) to coefficients of the f_j. Thus the dimension of $\mathbf{A}$ is n_j by the number of local mesh points, n_k. This matrix is constructed to reproduce the values of the planewaves as well as possible given the functions f_j. This means that standard least squares is to be used if $n_j < n_k$ (in which case the values on the mesh points will be reproduced approximately) and a constrained interpolation with a smoothness criterion [88, 147] may be used otherwise (in this case the values on the mesh points are reproduced exactly). Using least squares ($n_j < n_k$), $\mathbf{A}$ is

$$\mathbf{A} = \mathbf{C}^{-1}\mathbf{D}, \tag{6.17}$$

with

$$D_{ik} = f_i(\mathbf{r}_k), \tag{6.18}$$

and

$$C_{ij} = \sum_k f_i(\mathbf{r}_k) f_j(\mathbf{r}_k). \tag{6.19}$$

The matrix $\mathbf{A}$ as defined above depends on the locations, $\mathbf{r}_k$ of the FFT points relative to the center of the sphere, but does not depend on the values of the wavefunction on these points. This is also the case if constrained interpolation is used instead of least squares.

The essential ingredient in an iterative diagonalization is the ability to repeatedly compute $H\varphi$ for the wavefunctions of interest. At first sight it would seem that the contribution from the non-local pseudopotential, V^{NL}, may be calculated using $\mathbf{A}$ as follows: First an FFT might be used to transform φ to real space, and vectors Ψ consisting of the values of the φ on the local mesh points assembled. These then would be multiplied by $\mathbf{A}$ to yield the expansion coefficients of the f_j. One could then imagine operating the non-local pseudopotential on the f_j and the result accumulated on the mesh points. This then would be back transformed to reciprocal space. This procedure is, however, deficient for two reasons. First of all, it is not Hermetian as required by most iterative diagonalization schemes. Secondly, it is unable to couple the φ to strongly varying potentials as V^{NL} might be in certain cases. Specifically, while this situation is unlikely to arise in the context of planewave pseudopotential calculations (if V^{NL} is too strongly varying to be represented on the FFT mesh, the calculation is no doubt underconverged with respect to the planewave cutoff) it does arise in adapting the method to mixed basis and LAPW methods. These difficulties may be avoided by using the projector basis functions, f_j, explicitly.

First the matrix $\mathbf{T}$, defined by

$$\mathbf{T} = \mathbf{A}^{\mathrm{T}}\mathbf{V}\mathbf{A}, \tag{6.20}$$

is constructed. Here $\mathbf{V}$ is a square ($n_j \times n_j$) matrix consisting of the matrix elements of V^{NL} in the space of the f_j, and $\mathbf{A}^{\mathrm{T}}$ denotes the transpose of $\mathbf{A}$.

$$V_{ij} =< f_i|V^{NL}|f_j > . \tag{6.21}$$

The evaluation of these matrix elements is straightforward with a semi-local pseudopotential because of the angular momentum representation of the f_j. Then to the extent that the projector basis functions are flexible enough to mimic the planewave representation of the wavefunction inside the sphere,

$$< \varphi_m|V^{NL}|\varphi_n >\sim \Psi_m^\dagger \mathbf{T}\Psi_n, \tag{6.22}$$

where the φ_m are linear combinations of planewaves, the Ψ_m are vectors consisting of the values of the φ_m on the real space mesh points in the sphere, and V^{NL} is assumed to be zero outside r_c. The matrix $\mathbf{T}$ may then be used for the operation of V^{NL} in the determination of $H\varphi$.

Points to note are that: (1) $\mathbf{T}$ is manifestly Hermetian, (2) Computation of $\mathbf{T}\Psi$ does not yield the same values on the real space mesh as would be obtained by applying V^{NL} to the wavefunction and transforming to real space and (3) By using $\mathbf{T}\Psi$ to operate V^{NL} in real space, strongly varying potentials can be accommodated.

In order to illustrate points (2) and (3), let us consider the following (extreme) example. Suppose that φ is a constant, c. Thus $\Psi_k = c$. Also, suppose that V^{NL} is a delta function that is incommensurate with the FFT mesh, *i.e.* $V^{NL}(\mathbf{r}) = \delta(\mathbf{r} - \mathbf{r}_0)$ and therefore $V^{NL}(\mathbf{r}_k) = 0$ for all $\mathbf{r}_k$ because of the incommensurate delta function. Since in this case, V^{NL} is in fact local, we may consider first the usual real space procedure for operating a local potential, *i.e.* multiply on the real space FFT grid points. In our example, this yields zero because $V^{NL}(\mathbf{r}_k) = 0$. Similarly, if we transform to the space of the projector basis functions using $\mathbf{A}$, apply V^{NL} in this space (obtaining $c\delta(\mathbf{r} - \mathbf{r}_0)$) and then compute the values on the mesh points, we obtain zero. However, if we apply the matrix $\mathbf{T}$ we do not obtain zero. Rather, we note that we may write $\Psi_m^\dagger \mathbf{T}\Psi_n$ as $(\mathbf{A}\Psi_m)^\dagger \mathbf{V}(\mathbf{A}\Psi_n)$ and that $\mathbf{A}\Psi$ yields a coefficient $(4\pi)^{1/2}c$ for the constant projector basis function and zero for the other f_j. Further, in this case, the entries in $\mathbf{V}$ are given by $V_{ij} = f_i(\mathbf{r}_0)f_j(\mathbf{r}_0)$. This means, first of all, that the values of $\mathbf{T}\Psi$ are non-zero on the mesh points (and therefore not equal to $V^{NL}(\mathbf{r}_k)\varphi(\mathbf{r}_k)$) and that $\Psi_m^\dagger \mathbf{T}\Psi_n$ is equal to $\tilde{\varphi}_m^*(\mathbf{r}_0)\tilde{\varphi}_n(\mathbf{r}_0)$ with this choice of V^{NL}, where $\tilde{\varphi}(\mathbf{r})$ denotes the value of the interpolating combination of projector basis functions that represent $\varphi(\mathbf{r})$ between the mesh points.

In a nutshell, replacing V^{NL} by $\mathbf{T}$ yields correct matrix elements among wavefunctions that can be represented on a given FFT mesh, even if V^{NL} is too strongly varying for this mesh. This is because V^{NL} is never applied directly to planewaves in this formulation.

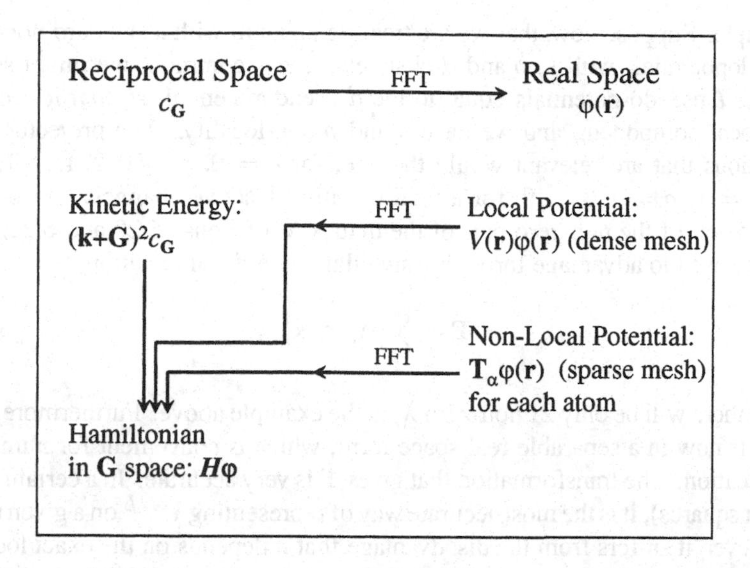

Figure 6.1. Computation of $H\varphi$ with a non-local pseudopotential and the projector basis method.

How then can this formalism be applied to construct a planewave pseudopotential code? The algorithm (Fig. 6.1) closely parallels a CP code with a local pseudopotential. The kinetic energy component is obtained in reciprocal space, where it is diagonal. Each wavefunction is then transformed in turn to real space using an FFT. The local component of the potential is applied directly on the FFT grid where it is diagonal. Finally, V^{NL} is applied by gathering the values of the wavefunction (with a phase factor for general **k**-points in the Brillouin zone) on the grid points into vectors Ψ (one per atom) and multiplying them by the corresponding **T** matrix, which is assumed to have been precalculated. The result of this operation and that of the local potential are then back transformed to reciprocal space and added to the result of the kinetic energy operation to obtain $H\varphi$.

Finally, we note that although it would seem at first sight that the storage and operation of the matrix **T** could be burdensome in practice, this is not the case. While, the dimension of **T** is $n_p \times n_p$, **T** actually contains far less information than this. This is because V^{NL} is diagonal in ℓ and only includes low ℓ components. Thus, what **T** really contains is information about the projection onto the low ℓ projector basis functions, and the matrix elements in this space. For example, if there are of order $n_p = 500$ mesh points in a given sphere, it may be convenient to use polynomials up to $p_{max} = 10$. This would yield, $n_j = 220$, which would be a convenient number for numerical stability in a least square fit. The corresponding dimension of **T** would be 500×500 in this example. Suppose now, that we are treating an atom with a standard non-local pseudopotential, with s, p and d channels. As is common, we might set the higher ℓ pseudo-potentials equal to the d pseudopotential, so that it becomes the local component, and we have s and p non-locality. The projector basis functions that are relevant would then be, for $\ell = 0$, $p = 0, 2, 4, ..., 10$ and for $\ell = 1$, $p = 1, 3, ..., 9$, for a total of only 21 active channels. This is the dimension of the non-zero part of the matrix $\mathbf{V}$ of Eqns. 6.20 and 6.21. This can be used to advantage through a singular value decomposition,

$$\mathbf{T} = \sum_i \lambda_i \, \mathbf{x}_i \, \mathbf{x}_i^{\dagger}, \tag{6.23}$$

Here there will be only 21 non-zero λ_i in the example above. Furthermore, Eqn. 6.23 is now in a separable real space form, which is convenient for numerical application. The transformation that gives $\mathbf{T}$ is very accurate. In a certain sense (least squares), it is the most accurate way of representing V^{NL} on a given mesh. However, it suffers from the disadvantage that it depends on the exact location of the atom relative to the mesh, and so it must be recomputed at every step in a dynamics run. As in a standard CP calculation, the computational cost of the operations in a planewave code using this method (Fig. 6.1) is dominated by the FFTs for large systems, yielding the same scaling. In particular, the cost of operating V^{NL} in real space is linear in the number of atoms. It is important to note that the operation of V^{NL} can be performed using an FFT mesh that is just adequate for representing a single wavefunction. This is in contrast to the operation of the local potential, which requires a grid capable of representing a product of two wavefunctions in order to avoid aliasing. Thus the FFT grid for operation of VNL may contain a factor of $2^3 = 8$ fewer points than that used for the local potential, although it may be advantageous to use an intermediate value instead for numerical stability.

Application to the LAPW Method

In order to apply the projector basis technique to the LAPW method, we consider a single atom for simplicity (the extension to many atoms is trivial, as will become evident) and write the potential, $V(\mathbf{r})$ and wavefunctions, $\varphi(\mathbf{r})$ as

$$V(\mathbf{r}) = [1 - \Theta(r - R_\alpha)]V_c(\mathbf{r}) + \Theta(r - R_\alpha)V^I(\mathbf{r}), \tag{6.24}$$

and

$$\varphi(\mathbf{r}) = [1 - \Theta(r - R_\alpha)]\varphi_c(\mathbf{r}) + \Theta(r - R_\alpha)\varphi^I(\mathbf{r}), \tag{6.25}$$

where V^I and φ^I are the interstitial planewave representations over all space (*i.e.* continued into the spheres) and $\Theta(x)$ is a step function that is zero for negative arguments and unity for positive arguments. We begin with the Hamiltonian given by the smooth interstitial Hamiltonian, $\tilde{H}^I = K.E. + V^I$ and the wavefunctions as linear combinations of planewaves, φ^I, and then compute the correction terms needed to convert this into the LAPW method in real space. To do this, the projector basis functions will be used as a local representation of the planewaves, and these will be used to obtain the matching conditions for the $A_{\ell m}$ and $B_{\ell m}$ as well as to operate the step functions. This yields two local matrices, $\mathbf{T}^{LAPW}$ and $\mathbf{O}^{LAPW}$ that are applied exactly as in the computation of V^{NL} described above.

First, let us consider the terms that need to be subtracted in order to account for the fact that the planewaves are prevented from extending into the sphere by the step function. The resulting matrices will be denoted $\mathbf{T}^{(1)}$ and $\mathbf{O}^{(1)}$ for the Hamiltonian and overlap, respectively. Exactly as in the planewave method, these are

$$\mathbf{T}^{(1)} = \mathbf{A}^{\mathrm{T}}\mathbf{H}^I\mathbf{A}, \tag{6.26}$$

and

$$\mathbf{O}^{(1)} = \mathbf{A}^{\mathrm{T}}\mathbf{R}^I\mathbf{A}, \tag{6.27}$$

where the $n_j \times n_j$ matrices $\mathbf{H}^I$ and $\mathbf{R}^I$ consist of the interstitial Hamiltonian and overlap matrix elements among the f_j. Since $\mathbf{T}^{(1)}$ and $\mathbf{O}^{(1)}$ are to be used to subtract the spurious sphere contribution, the integrations for these matrix elements are restricted to the interior of the sphere.

Next let us consider the terms, $\mathbf{T}^{(2)}$ and $\mathbf{O}^{(2)}$, for the Hamiltonian and overlap, respectively, that need to be added to account for the augmentation. At this stage, it is convenient to construct a matrix, $\mathbf{U}$, that maps from the space of projector basis functions, $\{f_j\}$ to the space of $A_{\ell m}$ and $B_{\ell m}$ coefficients. Thus the dimension of $\mathbf{U}$ is the total number of $A_{\ell m}$ and $B_{\ell m}$ coefficients by the number of projector basis functions, *i.e.* $2(\ell_{max}+1)^2 \times n_j$. The construction of $\mathbf{U}$ is straightforward because the f_j are already in an angular momentum representation, and their values are derivatives are known analytically on the sphere boundary (recall that they are just $r^p Y_{LM}(\mathbf{r})$, so the value is $R_\alpha^p/(4\pi)^{1/2}$ and the derivative is $pR_\alpha^{p-1}/(4\pi)^{1/2}$). Having done this, $\mathbf{T}^{(2)}$ and $\mathbf{O}^{(2)}$ can be built up as,

$$\mathbf{T}^{(2)} = \mathbf{A}^{\mathrm{T}}\mathbf{U}^{\mathrm{T}}\mathbf{H}^{(2)}\mathbf{U}\mathbf{A}, \tag{6.28}$$

and

$$\mathbf{O}^{(2)} = \mathbf{A}^{\mathrm{T}}\mathbf{U}^{\mathrm{T}}\mathbf{D}^{(2)}\mathbf{U}\mathbf{A}, \tag{6.29}$$

where $\mathbf{H}^{(2)}$ consists of the Hamiltonian matrix elements in the space of the radial functions, as given in Chapter 5, and the $\mathbf{D}^{(2)}$ are the corresponding overlap matrix elements. Here, $\mathbf{H}^{(2)}$ and $\mathbf{D}^{(2)}$ are to be evaluated over the interior of the sphere in question only.

Assembling the two contributions, above, we obtain

$$\mathbf{T}^{LAPW} = \mathbf{T}^{(2)} - \mathbf{T}^{(1)}, \tag{6.30}$$

and

$$\mathbf{O}^{LAPW} = \mathbf{O}^{(2)} - \mathbf{O}^{(1)}, \tag{6.31}$$

This transforms the LAPW method into a planewave method with a real space pseudopotential ($\mathbf{T}^{LAPW}$, one such matrix per atom) and a non-diagonal overlap, $\mathbf{O}^{LAPW}$ which is also computed in real space. The sphere charge density may also be constructed efficiently in real space using this framework. To do this, FFTs are used to transform the interstitial planewave wavefunctions to real space, the vectors Ψ are assembled, and then multiplied by $\mathbf{UA}$. This yields the $A_{\ell m}$ and $B_{\ell m}$ coefficients needed for the charge density, without requiring a sum over planewaves and atoms.

Despite the simple appearing form that results from this transformation, its application is considerably more complex than implementing a standard planewave based method. The reason is that underlying it all is the LAPW method, and so all the complexities in computing and manipulating the LAPW potential are retained. Secondly, although, as with the planewave method, the A matrices depend only on the atomic positions and therefore need be computed only once, the other matrices, *i.e.* $\mathbf{U}, \mathbf{H}^{I}, \mathbf{H}^{(2)}$ and $\mathbf{D}^{(2)}$ depend on the potential either directly, or indirectly through the dependence of u_ℓ and $\dot{u}_\ell$. This means that these matrices need to be recomputed each time the potential is updated, just as the projectors in the method of Goedecker and Maschke need to be updated. Furthermore, the gain from using a singular value decomposition of the $\mathbf{T}^{LAPW}$ and $\mathbf{O}^{LAPW}$ matrices is smaller than the corresponding gain in the case of a non-local pseudopotential method, since the $A_{\ell m}$ and $B_{\ell m}$ coefficients go up to high values of ℓ. While this projector approach to the LAPW method was implemented and tested [109], it has not been used in practice because the overhead that results from the repeated calculation of $\mathbf{T}^{LAPW}$ and $\mathbf{O}^{LAPW}$ is not made up for by better scaling for the system sizes that are commonly studied using the LAPW method.

6.4 Status and Outlook

Since the first edition of this book, the LAPW method and its extensions have emerged as a popular method for calculating of properties of materials from first principles. This owes much to the widespread availability of user-friendly codes, like WIEN2K [19]. Of equal significance is the fact that many properties, such as optical properties, core level spectroscopies and electric field gradients have been implemented in these codes. There are also a number of important advances in the basic LAPW methodology itself. These include the APW+LO method, the $p_{1/2}$ local orbital extension for accurate inclusion of spin-orbit, the LDA+U method, and non-collinear magnetism.

However, regardless of the generally better accuracy of the LAPW method, and the formal relationships between the LAPW method and pseudopotentials discussed in this book, planewave based approaches, especially with ultrasoft pseudopotentials, remain the method of choice for many problems, especially problems involving dynamics or complex structural relaxations. This is because Car-Parrinello-like algorithms have been very effectively exploited in pseudopotential codes, while only some aspects have been used in LAPW codes to date. Two serious problems that currently prevent LAPW structural relaxations from being as efficient as relaxations using ultrasoft pseudopotentials are (1) the ill-conditioning of the LAPW and especially APW+LO secular equations and (2) the difficulty in forming a good initial all-electron charge density in the LAPW representation after moving atoms.

The former problem slows the convergence of iterative diagonalizations, and mandates the use of exact diagonalization at least for a fraction of the steps. While various preconditioning schemes have been tried, an effective and computationally efficient updating scheme for the wavefunctions has yet to be found. The second problem compounds this difficulty, because more self-consistent iterations are required following each move in the absence of a good starting density. Typically, current codes construct the starting density after an atomic move by rigidly displacing the LAPW spheres, holding the interstitial expansion of the charge density fixed, and then adding a constant background charge to obtain overall neutrality. This could perhaps be improved by a decomposition of the density into overlapping atomic-like charges against a background. The atomic-like charges would then be moved with the atoms. The implementation could at least in principle be done using the same computational machinery as is used for overlapping atomic and core charges.

The connections between the LAPW method and planewave methods have been used to good advantage in making the LAPW method as efficient as it is. These include, for example, the use of Fourier transforms in constructing the charge densities from wavefunctions and in the construction of the long range Coulomb potential. However, the connections between these two families of methods have not yet been fully exploited. Besides the iterative wavefunction updating, discussed above, one may envision the increased use of real-space algorithms, the use of adaptive grids, and other improvements in future LAPW codes. These may eventually make the LAPW method the method of choice for large scale first principles molecular dynamics and structural relaxations, at least for systems where the high accuracy of the LAPW method is important.

References

[1] A. Aguayo, I.I. Mazin and D.J. Singh, Phys. Rev. Lett. **92**, 147201 (2004).

[2] O.K. Andersen, Phys. Rev. B **12**, 3060 (1975).

[3] V.I. Anisimov and O. Gunnarsson, Phys. Rev. B **43**, 7570 (1991).

[4] V.I. Anisimov, F. Aryasetiawan and A.I. Lichtenstein, J. Phys.: Condens. Matter **9**, 767 (1997).

[5] N.C. Bacalis, K. Blathras, P. Thomaides and D.A. Papaconstantopoulos, Phys. Rev. B **32**, 4849 (1985).

[6] G.B. Bachelet, D.R. Hamann and M. Schluter, Phys. Rev. B **26**, 4199 (1982).

[7] A. Baldereschi, Phys. Rev. B**7**, 5212 (1973).

[8] S. Baroni, P. Giannozzi and A. Testa, Phys. Rev. Lett. **58**, 1861 (1987).

[9] S. Baroni, S. de Gironcoli and A. Dal Corso, Rev. Mod. Phys. **73**, 515 (2001).

[10] S. Baroni, A. Dal Corso, S. de Gironcoli, P. Giannozzi, C. Cavazzoni, G. Ballabio, S. Scandolo, G. Chiarotti, P. Focher, A. Pasquarello, K. Laasonen, A. Trave, R. Car, N. Marzari, and A. Kokalj, http://www.pwscf.org.

[11] A.D. Becke, J. Chem. Phys. **84**, 4524 (1986).

[12] A.D. Becke, Phys. Rev. A **38**, 3098 (1988).

[13] A.D. Becke, J. Chem. Phys. **97**, 9173 (1992).

[14] A.D. Becke, J. Chem. Phys. **107**, 8554 (1997).

[15] P. Bendt and A. Zunger, Phys. Rev. B **26**, 3144 (1982).

[16] P. Bendt and A. Zunger, Phys. Rev. Lett. **50**, 1684 (1983).

[17] P. Blaha, K. Schwarz, P. Sorantin and S.B. Trickey, Comp. Phys. Commun. **59**, 399 (1990).

[18] P. Blaha, D.J. Singh, P.I. Sorantin and K. Schwarz, Phys. Rev. B **46**, 1321 (1992).

[19] P. Blaha, K. Schwarz G.K.H. Madsen, D. Kvasnicka, and J. Luitz, *WIEN2K, An Augmented Plane Wave + Local Orbitals Program for for Calculating Crystal Properties* (K. Schwarz, Techn. Universitat Wien, Austria, 2001), ISBN 3-9501031-1-2. http://www.wien2k.at

[20] P.E. Blöchl, Phys. Rev. B **41**, 5414 (1990).

[21] P.E. Blöchl, unpublished, presented at the APS March Meeting, Seattle, USA (1993).

[22] P.E. Blöchl, Phys. Rev. B **50**, 17953 (1994).

[23] D.M. Brink and G.R. Satchler, *Angular Momentum* (Clarendon, Oxford, 1968).

[24] K.D. Brommer, M. Needels, B.E. Larson and J.D. Joannopoulos, Phys. Rev. Lett. **68**, 1355.

[25] H. Bross, Phys. Kondens. Mater **3**, 119 (1964).

[26] H. Bross, G. Bohn, G. Meister, W. Schube and H. Stohr, Phys. Rev. B **2**, 3098 (1970).

[27] J.Q. Broughton and F. Khan, Phys. Rev. B **40**, 12098 (1989).

[28] C.G. Broyden, Math. Comp. **19**, 577 (1965).

[29] J. Callaway and N.H. March, in *Solid State Physics*, H. Ehrenreich and D. Turnbull, eds. **38**, 135 (Academic Press, New York, 1984).

[30] R. Car and M. Parrinello, Phys. Rev. Lett. **55**, 2471 (1985).

[31] P. Carrier and S.-H. Wei, Phys. Rev. B **70**, 035212 (2004).

[32] D.M. Ceperley and B.J. Alder, Phys. Rev. Lett. **45**, 566 (1980).

[33] D.J. Chadi and M.L. Cohen, Phys. Rev. B **8**, 5747 (1973).

[34] J.-H. Cho and M.-H. Kang, Phys. Rev. B **52**, 9159 (1995).

[35] M.L. Cohen, Phys. Rep. **110**, 293 (1984).

[36] R.E. Cohen, unpublished (1989).

[37] R.E. Cohen, W.E. Pickett and H. Krakauer, Phys. Rev. Lett. **674**, 2575 (1990).

[38] P. de Ciccio, Phys. Rev. **153**, 931 (1967).

[39] P.H. Dederichs and R. Zeller, Phys. Rev. B **28**, 5462 (1983).

[40] J.T. Devreese and P. Van Camp, *Electronic Structure, Dynamics and Quantum Structural Properties of Matter* (Plenum, New York, 1985).

[41] P.A.M. Dirac, Proc. Roy. Soc. (London) **123**, 714 (1929).

[42] R.M. Dreizler and J. da Provincia, *Density Functional Methods in Physics* (Plenum, New York, 1985).

[43] N. Elyashar and D.D. Koelling, Phys. Rev. B **13**, 5362 (1976).

[44] M. Ernzerhof, J.P. Perdew and K. Burke, *Density Functionals: Where do they come from, why do they work?*, in *Topics in Current Chemistry*, vol. 180, R.F. Nalewajski ed. (Springer, Berlin, 1996).

[45] H. Eschrig and W.E. Pickett, Solid State Commun. **118**, 123 (2001).

[46] R.P. Feynmann, Phys. Rev. **56**, 340 (1939).

[47] G. Galli, R.M. Martin, R. Car and M. Parrinello, Phys. Rev. Lett. **63**, 988 (1989).

[48] A. Garcia, C. Elsasser, J. Zhu, S.G. Louie and M.L. Cohen, Phys. Rev. B **46**, 9829 (1992).

[49] P. Giannozzi, S. de Gironcoli, P. Pavone and S. Baroni, Phys. Rev. B **43**, 7231 (1991).

[50] G. Gilat, J. Comp. Phys. **10**, 432 (1972).

[51] S. de Gironcoli, Phys. Rev. B **51**, 6773 (1995).

[52] S. Goedecker and K. Maschke, Phys. Rev. B **42**, 8858 (1990).

[53] S. Goedecker and K. Maschke, Phys. Rev. B **45**, 1597 (1992).

[54] S. Goedecker and K. Maschke, Phys. Rev. A **45**, 88 (1992).

[55] S. Goedecker, Phys. Rev. B **47**, 9881 (1993).

[56] S. Goedecker, Rev. Mod. Phys. **71**, 1085 (1999).

[57] X. Gonze and J.-P. Vigneron, Phys. Rev. B **39**, 13120 (1989).

[58] X. Gonze, Phys. Rev. A **52**, 1086 (1995).

[59] X. Gonze, Phys. Rev. A **52**, 1096 (1995).

[60] X. Gonze, J.-M. Beuken, R. Caracas, F. Detraux, M. Fuchs, G.-M. Rignanese, L. Sindic, M. Verstraete, G. Zerah, F. Jollet, M. Torrent, A. Roy, M. Mikami, Ph. Ghosez, J.-Y. Raty and D.C. Allan, Computational Materials Science **25**, 478-492 (2002). http://www.abinit.org

[61] O. Gunnarsson and B.I. Lundqvist, Phys. Rev. B **13**, 4274 (1976).

[62] O. Gunnarsson, M. Jonson and B.I. Lundqvist, Solid State Commun. **24**, 765 (1977).

[63] O. Gunnarsson, M. Jonson and B.I. Lundqvist, Phys. Rev. B **20**, 3136 (1979).

[64] O. Gunnarsson and R.O. Jones, Phys. Scr. **21**, 394 (1980).

[65] J. Haglund, Phys. Rev. B **47**, 566 (1993).

[66] D.R. Hamann, Phys. Rev. Lett. **42**, 662 (1979).

[67] D.R. Hamann, M. Schluter and C. Chiang, Phys. Rev. Lett. **43**, 1494 (1979).

[68] D.R. Hamann, L.F. Mattheiss and H.S. Greenside, Phys. Rev. B **24**, 6151 (1981).

[69] L. Hedin and B.I. Lundqvist, J. Phys. C **4**, 2064 (1971).

[70] V. Heine and M.J.G. Lee, Phys. Rev. Lett. **27**, 811 (1971).

[71] H. Hellmann, *Einfuhrung in die Quantenchemie*, pg. 285 (Deuieke, Leipzig, 1937).

[72] C. Herring, Phys. Rev. **57**, 1169 (1940).

[73] H. Hohenberg and W. Kohn, Phys. Rev. **136** B864 (1964).

[74] M.S. Hybertsen and S.G. Louie, Solid State Commun. **51**, 451 (1984).

[75] J. Ihm, A. Zunger and M.L. Cohen, J. Phys. C **12**, 4409 (1979).

[76] H.J.F. Jansen and A.J. Freeman, Phys. Rev. B **30**, 561 (1984).

[77] O. Jepsen and O.K. Andersen, Solid State Commun. **9**, 1763 (1977).

[78] O. Jepsen, J. Madsen and O.K. Andersen, Phys. Rev. B **18**, 605 (1978).

[79] D.D. Johnson, Phys. Rev. B **38**, 12807 (1988).

[80] G.P. Kerker, J. Phys. C **13**, L189 (1980).

[81] R.D. King-Smith, M.C. Payne and J.S. Lin, Phys. Rev. B **44**, 13063 (1991).

[82] S. Kirkpatrick, C.D. Gelatt Jr. and M.P. Vecchi, Science **220**, 671 (1983).

[83] L. Kleinman and D.M. Bylander, Phys. Rev. Lett. **48**, 1425 (1982).

[84] D.D. Koelling, Phys. Rev. **188**, 1049 (1969).

[85] D.D. Koelling, Phys. Rev. B **2**, 290 (1970).

[86] D.D. Koelling and G.O. Arbman, J. Phys. F **5**, 2041 (1975).

[87] D.D. Koelling and B.N. Harmon, J. Phys. C **10**, 3107 (1977).

[88] D.D. Koelling and J.H. Wood, J. Comp. Phys. **67**, 253 (1986).

[89] K. Koepernik and H. Eschrig, Phys. Rev. B **59**, 1743 (1999).

[90] W. Kohn and L.J. Sham, Phys. Rev. **140**, A1133 (1965).

[91] M. Korling and J. Haglund, Phys. Rev. B **45**, 13293 (1992).

[92] H. Krakauer, M. Posternak and A.J. Freeman, Phys. Rev. B **19**, 1706 (1979).

[93] G. Kresse, J. Hafner and R.J. Needs, J. Phys. Condens. Matter **4**, 7451 (1992).

[94] J. Kubler, K.-H. Hock, J. Sticht and A.R. Williams, J. Phys. F **18**, 469 (1988).

[95] J. Kunes, P. Novak, M. Divis and P.M. Oppeneer, Phys. Rev. B **63**, 205111 (2001).

[96] J. Kunes, P. Novak, R. Schmid, P. Blaha and K. Schwarz, Phys. Rev. B **64**, 153102 (2001).

[97] Ph. Kurz, F. Forster, L. Nordström, G. Bihlmayer and S. Blugel, Phys. Rev. B **69**, 024415 (2004).

[98] K. Laasonen, R. Car, C. Lee and D. Vanderbilt, Phys. Rev. B **43**, 6796 (1991).

[99] K. Laasonen, A. Pasquarello, R. Car, C. Lee and D. Vanderbilt, Phys. Rev. B **47**, 10142 (1993).

[100] D.C. Langreth and J.P. Perdew, Solid State Commun. **17**, 1425 (1975).

[101] P. Larson, I.I. Mazin and D.J. Singh, Phys. Rev. B **69**, 064429 (2004).

[102] D.C. Langreth and M.J. Mehl, Phys. Rev. Lett. **47**, 446 (1981).

[103] M.J.G. Lee and V. Heine, Phys. Rev. B **5**, 3839 (1972).

[104] G. Lehmann, P. Rennert, M. Taut and H. Wonn, Phys. Status Solidi **37**, K27 (1970).

[105] G. Lehmann and M. Taut, Phys. Status Solidi **54**, 469 (1972).

[106] X.P. Li, R.W. Nunes and D. Vanderbilt, Phys. Rev. B **47**, 10891 (1993).

[107] D. Liberman, J.T. Waber and D.T. Cromer, Phys. Rev. **137**, A27 (1965).

[108] A.I. Liechtenstein, V.I. Anisimov and J. Zaanen, Phys. Rev. B **52**, 5467 (1995).

[109] A.Y. Liu, D.J. Singh and H. Krakauer, Phys. Rev. B **49**, 17424 (1994).

[110] S.G. Louie, S. Froyen and M.L. Cohen, Phys. Rev. B **26**, 1738 (1982).

[111] T.L. Loukes, *The Augmented-Plane-Wave Method* (Benjamin, New York, 1967).

[112] S. Lundqvist and N.H. March, eds. *Theory of the Inhomogeneous Electron Gas* (Plenum, New York, 1983).

[113] A.H. MacDonald, W.E. Pickett and D.D. Koelling, J. Phys. C **13**, 2675 (1980).

[114] P.M. Marcus, Int. J. Quantum Chem. **1**, 567 (1967).

[115] L.F. Mattheiss and D.R. Hamann, Phys. Rev. B **33**, 823 (1986).

[116] I.I. Mazin and D.J. Singh, Phys. Rev. B **69**, 020402 (2004).

[117] A.D. McLaren, Math. Comp. **17**, 361 (1963).

[118] H.J. Monkhorst and J.D. Pack, Phys. Rev. B **13**, 5188 (1976).

[119] H.J. Monkhorst and J.D. Pack, Phys. Rev. B **16**, 1748 (1977).

[120] V.L. Moruzzi, J.F. Janak and A.R. Williams, *Calculated Electronic Properties of Metals* (Pergamon, New York, 1978).

[121] V.L. Moruzzi, Phys. Rev. Lett. **57**, 2211 (1986).

[122] V.L. Moruzzi, P.M. Marcus, K. Schwarz and P. Mohn, Phys. Rev. B **34**, 1784 (1986).

[123] V.L. Moruzzi and P.M. Marcus, J. Appl. Phys. **64**, 5598 (1988).

[124] V.L. Moruzzi, P.M. Marcus and P.C. Pattnaik, Phys. Rev. B **37**, 8003 (1988).

[125] V.L. Moruzzi, Physica B **161**, 99 (1989).

[126] V.L. Moruzzi and P.M. Marcus, Solid State Commun. **71**, 203 (1989).

[127] V.L. Moruzzi and P.M. Marcus, Phys. Rev. B **39**, 471 (1989).

[128] V.L. Moruzzi, P.M. Marcus and J. Kubler, Phys. Rev. B **39**, 6957 (1989).

[129] V.L. Moruzzi, Phys. Rev. B **41**, 6939 (1990).

[130] V.L. Moruzzi and P.M. Marcus, Phys. Rev. B **42**, 8361 (1990).

[131] V.L. Moruzzi and P.M. Marcus, Phys. Rev. B **42**, 10322 (1990).

[132] L. Nordström and D. J. Singh, Phys. Rev. Lett. **76**, 4420 (1996).

[133] L. Nordström and A. Mavromaras, Europhys. Lett. **49**, 775 (2000).

[134] L. Nordström, J.M. Wills, P.H. Andersson, P. Söderlind and O. Eriksson, Phys. Rev. B **63**, 035103 (2000).

[135] R.W. Nunes and D. Vanderbilt, Phys. Rev. B **50**, 17611 (1994).

[136] T. Ono and K. Hirose, Phys. Rev. Lett. **82**, 5016 (1999).

[137] R.G. Parr and W. Yang, *Density Functional Theory of Atoms and Molecules* (Oxford University Press, New York, 1989).

[138] M.C. Payne, J.D. Joannopoulos, D.C. Allan, M.P. Teter and D.H. Vanderbilt, Phys. Rev. Lett. **56**, 2656 (1986).

[139] M.C. Payne, M.P. Teter, D.C. Allan, T.A. Arias and J.D. Joannopoulos, Rev. Mod. Phys. **64**, 1045 (1992).

[140] J.P. Perdew and W. Yue, Phys. Rev. B **33**, 8800 (1986).

[141] J.P. Perdew, J.A. Chevary, S.H. Vosko, K.A. Jackson, M.R. Pederson, D.J. Singh and C. Fiolhais, Phys. Rev. B **46**, 6671 (1992).

[142] J.P. Perdew, K. Burke and M. Ernzerhof, Phys. Rev. Lett. **77**, 3865 (1996).

[143] J.P. Perdew and Y. Wang, Phys. Rev. B **46**, 12947 (1992).

[144] J. Petru and L. Smrcka, Czech J. Phys. B **35**, 62 (1985).

[145] A.G. Petukhov, I.I. Mazin, L. Chioncel and A.I. Lichtenstein, Phys. Rev. B **67**, 153106 (2003).

[146] J.C. Phillips and L. Kleinman, Phys. Rev. **116**, 287 (1959).

[147] W.E. Pickett, H. Krakauer and P.B. Allen, Phys. Rev. B **38**, 2721 (1988).

[148] W.E. Pickett, Comp. Phys. Rep. **9**, 115 (1989).

[149] D. Porezag, M.R. Pederson and A.Y. Liu, Phys. Rev. B **60**, 14132 (1999).

[150] W.H. Press, B.P. Flannery, S.A. Teukolsky and W.T. Vetterling, *Numerical Recipes* (Cambridge University Press, Cambridge, 1986).

[151] P. Pulay, Mol. Phys. **17**, 197 (1969).

[152] A.M. Rappe, K.M. Rabe, E. Kaxiras and J.D. Joannopoulos, Phys. Rev. B **41** 1227 (1990).

[153] J. Rath and A.F. Freeman, Phys. Rev. B **11**, 2109 (1975).

[154] F. Rosicky, P. Weinberger and F. Mark, J. Phys. B **9**, 2971 (1976).

[155] L. Sandratskii and P. Guletskii, J. Phys. F **16**, L43 (1986).

[156] L.M. Sandratskii, Adv. Phys. **47**, 91 (1988).

[157] L.M. Sandratskii, Phys. Rev. B **64**, 134402 (2001).

[158] K. Schwarz and P. Mohn, J. Phys. F **14**, L129 (1984).

[159] H. Schlosser and P.M. Marcus, Phys. Rev. **131**, 2529 (1963).

[160] D.J. Shaughnessy, G.R. Evans and M.I. Darby, J. Phys. F **17**, 1671 (1987).

[161] R.W. Shaw, Jr. and W.A. Harrison, Phys. Rev. **163**, 604 (1967).

[162] A.B. Schick, A.I. Liechtenstein and W.E. Pickett, Phys. Rev. B **60**, 10763 (1999).

[163] E.L. Shirley, D.C. Allan, R.M. Martin and J.D. Joannopoulos, Phys. Rev. B **40**, 3652 (1989).

[164] D. Singh, H. Krakauer and C.S. Wang, Phys. Rev. B **34**, 8391 (1986).

[165] D. Singh, Phys. Rev. B **40**, 5428 (1989).

[166] D. Singh, Phys. Rev. B **43**, 6388 (1991).

[167] D.J. Singh, Phys. Rev. B **44**, 7451 (1991).

[168] D. Singh and H. Krakauer, Phys. Rev. B **43**, 1441 (1991).

[169] D.J. Singh and J. Ashkenazi, Phys. Rev. B **46**, 11570 (1992).

[170] D.J. Singh, H. Krakauer, C. Haas and A.Y. Liu, Phys. Rev. B **46**, 10365 (1992).

[171] D.J. Singh, K. Schwarz and P. Blaha, Phys. Rev. B **46**, 5849 (1992).

[172] D.J. Singh, H. Krakauer, C. Haas and W.E. Pickett, Nature **365**, 39 (1993).

[173] D.J. Singh, *Planewaves Pseudopotentials and the LAPW Method* (Kluwer, Boston, 1994).

[174] E. Sjöstedt, L. Nordström and D.J. Singh, Solid State Commun. **114**, 15 (2000).

[175] E. Sjöstedt, and L. Nordström, Phys. Rev. B **66** 014447 (2002).

[176] J.C. Slater, Phys. Rev. **51**, 846 (1937).

[177] J.C. Slater, Phys. Rev. **81**, 385 (1951).

[178] J.C. Slater, Advances in Quantum Chemistry **1**, 35 (1964).

[179] J.C. Slater, *The Self-Consistent Field for Molecules and Solids* (McGraw-Hill, New York, 1974).

[180] L. Smrcka, Czech J. Phys. B **34**, 694 (1984).

[181] J.M. Soler and A.R. Williams, Phys. Rev. B **40**, 1560 (1989).

[182] J.M. Soler and A.R. Williams, Phys. Rev. B **42**, 9728 (1990).

[183] J.M. Soler and A.R. Williams, Phys. Rev. B **47**, 6784 (1993).

[184] J.M. Soler, E. Artacho, J.D. Gale, A. Garcia, J. Junquera, P. Ordejon and D. Sanchez-Portal, J. Phys. Condens. Matter **14**, 2745 (2002).

[185] I.V. Solovyev, A.I. Liechtenstein and K. Terakura, Phys. Rev. Lett. **80**, 5758 (1998).

[186] G.P. Srivastava, J. Phys. A **17**, L317 (1984).

[187] T. Starkloff and J.D. Joannopoulos, Phys. Rev. B **16**, 5212 (1977).

[188] R. Sternheimer, Phys. Rev. **84**, 244 (1951).

[189] I. Stich, M.C. Payne, R.D. King-Smith, J.S. Lin and L.J. Clarke, Phys. Rev. Lett. **68**, 1351 (1992).

[190] T. Takeda, J. Phys. F **9**, 815 (1979).

[191] T. Takeda and J. Kubler, J. Phys. F **9**, 661 (1979).

[192] W.C. Topp and J.J. Hopfi eld, Phys. Rev. B **7**, 1295 (1974).

[193] N. Troullier and J.L. Martins, Phys. Rev. B **43**, 1993 (1991).

[194] N. Troullier and J.L. Martins, Phys. Rev. B **43**, 8861 (1991).

[195] D. Vanderbilt and S.G. Louie, Phys. Rev. B **30**, 6118 (1984).

[196] D. Vanderbilt, Phys. Rev. B **32**, 8412 (1985).

[197] D. Vanderbilt, Phys. Rev. B **41**, 7892 (1990).

[198] U. von Barth and L. Hedin, J. Phys. C **5**, 1629 (1972).

[199] C.-Z. Wang, R.C. Yu and H. Krakauer, Phys. Rev. B **53**, 5430 (1996).

[200] C.-Z. Wang, R. Yu and H. Krakauer, Phys. Rev. B **54**, 11161 (1996).

[201] C.Z. Wang, R. Yu and H. Krakauer, Phys. Rev. B **59**, 9278 (1999).

[202] S.-H. Wei and H. Krakauer, Phys. Rev. Lett. **55**, 1200 (1985).

[203] S.-H. Wei, H. Krakauer and M. Weinert, Phys. Rev. B **32**, 7792 (1985).

[204] M. Weinert, J. Math. Phys. **22**, 2433 (1981).

[205] A.R. Williams, V. Moruzzi, J. Kubler and K. Schwarz, Bull. Am. Phys. Soc. (APS March Meeting) **29**, 278 (1984).

[206] A.R. Williams and J. Soler, Bull. Am. Phys. Soc. (APS March Meeting) **32**, 562 (1987).

[207] E. Wimmer, H. Krakauer, M. Weinert and A.J. Freeman, Phys. Rev. B **24**, 864 (1981).

[208] E. Wimmer, H. Krakauer and A.J. Freeman, Adv. Electronics Electron Phys. **65**, 337 (1985).

[209] J.H. Wood and A.M. Boring, Phys. Rev. B **18**, 2701 (1978).

[210] D.M. Wood and A. Zunger, J. Phys. A **18**, 1343 (1985).

[211] R. Yu, D. Singh and H. Krakauer, Phys. Rev. B **43**, 6411 (1991).

[212] R. Yu and H. Krakauer, Phys. Rev. B **49**, 4467 (1994).

[213] R. Yu and H. Krakauer, Phys. Rev. Lett. **74**, 4067 (1995).

[214] J.M. Ziman, in *Solid State Physics*, H. Ehrenreich, F. Seitz and D. Turnbull, eds. **26** (Academic Press, New York, 1971).

[215] A. Zunger and A.J. Freeman, Phys. Rev. B **15**, 5049 (1977).

[216] A. Zunger and A.J. Freeman, Phys. Rev. B **16**, 906 (1977).

[217] A. Zunger and A.J. Freeman, Phys. Rev. B **16**, 2901 (1977).

Index

$p_{1/2}$ local orbital, 105, 106

aliasing, 69, 118
angular momentum cutoff, choosing, 62
APW method, 43–45, 47, 88, 109, 111
APW+LO method, 48, 49, 88, 89, 96, 106, 111, 112, 121
asymptote problem, 45–47
atomic orbitals, 12
augmentation, 43, 44, 46, 47, 50, 62, 63, 66, 67, 72, 88, 96, 102

Bachelet, Hamann, Schluter pseudopotential, 28–30
Baldereschi point, 75
basis sets, 10, 11, 15
Bloch's theorem, 24
Bloch's theory, 10
Brillouin zone sampling, 10, 73, 74
Broyden's method, 14, 90–92

Car-Parrinello method, 13, 15, 23, 36–41, 90, 107, 108, 118, 121
charge density synthesis, 78–81
core correction, 30

density functional perturbation theory, 99–102
density functional theory, 5–8, 20
density matrix, 107
discrete inversion in iterative subspace (DIIS), 41, 106
double counting, 20
Dzyaloshinskii-Moriya interaction, 16

energy parameters, 43, 45, 47–52, 63, 64, 83, 85, 86
exchange-correlation functional, 7, 8, 10, 13
exchange-correlation potential, 60–62

fast Fourier transform, 39, 61, 80, 83, 113–118, 120
Fermi energy, 9, 51, 73, 74, 78, 93, 109
Fermi liquid theory, 25
fi xed spin moment method, 18, 92–94
force calculation, 95–99
frozen augmentation approximation, 98, 102
frozen core approximation, 26

Gaunt coeffi cients, 72, 73, 81
general potential, 45, 46
generalized gradient approximation, 5, 7–9, 19, 60
ghost bands, 50

Hamiltonian synthesis, 67–73
Hartree energy, 7, 17, 25, 57, 94
Hellman-Feynmann force, 97
Hellmann-Feynman force, 96

incomplete basis force, 96–99
iterative diagonalization, 15, 16, 39, 41, 106, 112, 115, 121

Kerker pseudopotential, 30
kinetic energy, 7–9, 11–13, 24, 44, 64, 68, 70, 89, 99, 102, 111, 117
Kleinman-Bylander transformation, 32
Kohn-Sham equation, 9–11, 101

Lagrangian, 36, 37
lattice harmonics, 44, 55–57, 62, 78, 80, 81, 83
LDA+U method, 20, 21, 121
least squares, 62, 114
linear response, 99–102
linearization, 46, 47, 63, 83
local density approximation, 5–8, 21, 60
local orbitals, 48, 51, 86–89, 96
local spin density approximation, 19, 95
logarithmic derivative, 28, 29, 46
logarithmic mesh, 57

Madelung potential, 95
modifi ed APW method, 46
Mott insulator, 20
muffi n-tin approximation, 44, 46
muffi n-tin orbitals, 12
multipole expansion, 58, 59

non-collinear magnetism, 6, 16, 18–20, 61, 90, 121
norm-conservation, 27, 28, 33

order N methods, 107
orthogonalized planewave method, 26
overlap synthesis, 67–71
overlapped atomic charges, 81–83

pair correlation function, 7, 8
PAW method, 108
Phillips-Kleinman pseudopotential, 26, 27, 109
Poisson equation, 57–60
preconditioning, 39, 41, 106, 112, 121
projector basis, 114–119
pseudo-charge, 83
pseudo-charge method, 58–60

radial function, 43, 44, 46–48, 62–66, 71, 85, 88, 102
reanalyzed potential, 70
relativity, 64–66
representative atom, 57, 67, 80, 81
residuals, 40, 41, 106

scalar relativistic approximation, 65, 102
second variational method, 12, 102–105
self-consistency, 10, 13, 14, 17
semi-core state, 50–52, 83–86
semi-local pseudopotential, 27, 28, 30, 31
separable pseudopotential, 32
singular value decomposition, 118, 120
SLAPW method, 48, 86, 87
special points method, 74–78
sphere radii, choosing, 49, 51, 52
spin density functional theory, 17, 18
spin spirals, 16, 20
spin-orbit, 12, 16, 20, 64, 65, 94, 102–106, 121
stars, 44, 54, 55, 78, 79
steepest descent, 41
step function, 68–70, 119
Sternheimer equation, 100, 101
straight mixing, 14, 90
structure factor, 31, 67, 83, 88

temperature broadening, 74, 78, 79
tetrahedron method, 73
total energy, 6, 7, 10, 11, 13, 17, 18, 23, 33, 36, 38, 50, 94, 95
transferability, 27–30, 33, 111
Troullier-Martins pseudopotential, 29

ultrasoft pseudopotential, 33–36, 107, 108, 121

variational principle, 9, 10, 13, 17, 34, 50, 95

warped muffi n-tin potential, 45
weighted density approximation, 7, 8